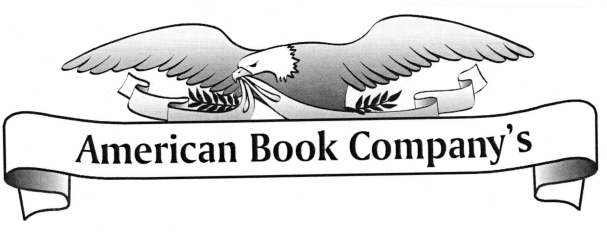

MASTERING THE GEORGIA
4th GRADE CRCT

IN

MATHEMATICS

Developed to the new Georgia Performance Standards!

ERICA DAY

COLLEEN PINTOZZI

MARY REAGAN

AMERICAN BOOK COMPANY

P. O. BOX 2638

WOODSTOCK, GEORGIA 30188-1383

TOLL FREE 1 (888) 264-5877   PHONE (770) 928-2834

FAX (770) 928-7483

WEB SITE: www.americanbookcompany.com

Acknowledgements

In preparing this book, we would like to acknowledge Mary Stoddard for her contributions in editing and developing graphics, and Trisha Paster for her contributions in editing and writing for this book. We would also like to thank our many students whose needs and questions inspired us to write this text.

Contents

Contents

Preface

Passing the Georgia 4th Grade CRCT in Mathematics will help you review and learn important concepts and skills related to elementary school mathematics. First, take the Diagnostic Test beginning on page 1 of the book. To help identify which areas are of greater challenge for you, complete the evaluation chart with your instructor in order to help you identify the chapters which require your careful attention. When you have finished your review of all of the material your teacher assigns, take the progress tests to evaluate your understanding of the material presented in this book. **The materials in this book are based on the Georgia Performance Standards including the content descriptions for mathematics, which are published by the Georgia Department of Education. The complete list of standards is located in the Answer Key. Each question in the Diagnostic and Practice Tests is referenced to the standard, as is the beginning of each chapter.**

This book contains several sections. These sections are as follows: 1) A Diagnostic Test; 2) Chapters that teach the concepts and skills for ***Passing the Georgia 4th Grade CRCT in Mathematics***; and 3) Two Practice Tests. Answers to the tests and exercises are in a separate manual.

ABOUT THE AUTHORS

Erica Day has a Bachelor of Science Degree in Mathematics and is working on a Master of Science Degree in Mathematics. She graduated with high honors from Kennesaw State University in Kennesaw, Georgia. She has also tutored all levels of mathematics, ranging from high school algebra and geometry to university-level statistics, calculus, and linear algebra. She is currently writing and editing mathematics books for American Book Company, where she has coauthored numerous books, such as ***Passing the Georgia Algebra I End of Course***, ***Passing the Georgia High School Graduation Test in Mathematics***, and ***Passing the New Jersey HSPA in Mathematics***, to help students pass graduation and end of course/grade exams.

Colleen Pintozzi has taught mathematics at the middle school, junior high, senior high, and adult level for 22 years. She holds a B.S. degree from Wright State University in Dayton, Ohio and has done graduate work at Wright State University, Duke University, and the University of North Carolina at Chapel Hill. She is the author of many mathematics books including such best-sellers as *Basics Made Easy: Mathematics Review, Passing the New Alabama Graduation Exam in Mathematics, Passing the Louisiana LEAP 21 GEE, Passing the Indiana ISTEP+ GQE in Mathematics, Passing the Minnesota Basic Standards Test in Mathematics,* and *Passing the Nevada High School Proficiency Exam in Mathematics.*

Mary Reagan

Diagnostic Test

Part One

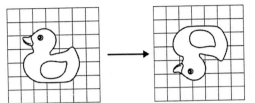

1. What degree of rotation does the figure above represent?

   A. 90°
   B. 120°
   C. 180°
   D. 360°

M4M2b

2. Which digit is the thousands place in 456,789?

   A. 7
   B. 6
   C. 5
   D. 4

M4N1a

3. A picture framer cuts each corner of a frame to a perfect 45° angle to make each corner fit tight. Which angle below most closely resembles a 45° angle?

   A.

   B.

   C.

   D.

M4M2a

4. Round 1,547 to the nearest hundred.

   A. 1,500
   B. 1,540
   C. 1,547
   D. 1,507

M4N2a

5. Find the sum. $45.85 + 62.37$.

   A. 107.52
   B. 106.22
   C. 105.27
   D. 108.22

M4N5c

6. Which lines are perpendicular?

   A.

   B.

   C.

   D.

M4G1b

7. Which ordered pair represents the point on the coordinate plane below?

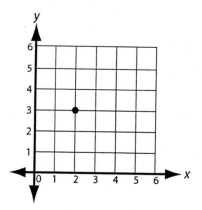

A. $(2, 3)$

B. $(3, 2)$

C. $(2, 2)$

D. $(3, 0)$

M4G3b

8. What is the total number of pounds of potatoes + chili peppers + bok choy?

Mrs. Eatyore's Vegetable Garden Yield

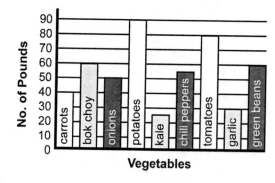

A. 200

B. 195

C. 205

D. 190

M4D1c

9. Which triangle below is an isosceles triangle?

A.

B.

C.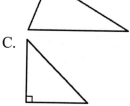

D.

M4G1a

10. What is twelve thousand, nine hundred sixty-three in standard form?

A. 12,639

B. 12,963

C. 69,123

D. 12,369

M4N1b

11. Find the product.

$234 \times 12 =$

A. 2,808

B. 2,340

C. 2,408

D. 2,342

M4N3

12. Solve.

$2 \times (3 + 6) - 7 \times 2 =$

A. -2

B. 6

C. 4

D. 13

M4N7b

13. How many flowers should be represented for spring? (Each symbol represents $10°$.)

Average Temperature of Seasons

Temperature	Season
70°	spring
95°	summer
75°	fall
60°	winter

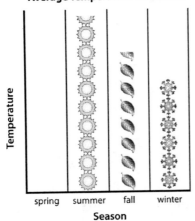

Average Temperature of Seasons

A. $7\frac{1}{2}$
B. 7
C. 9
D. 6

M4D1a

14. Round the following decimal to the nearest whole number.

24,518.47

A. 24,519
B. 24,518
C. 24,517
D. 24,520

M4N2c

15. Solve the following division problem.

$$172 \div 5 =$$

A. 34 r2
B. 85 r2
C. 62 r1
D. 34 r5

M4N4b

16. What is the measure of the angle?

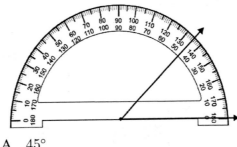

A. 45°
B. 55°
C. 135°
D. 145°

M4M2a

17. Which of the figures below has at least 3 perpendicular sides?

A.

B.

C.

D.

M4G1c

18. Identify the correct operation.

$$36 \triangle 6 = 6$$

A. $\times$

B. $-$

C. $\div$

D. $+$

19. How many more kids took the bus to school than bicycled to school?

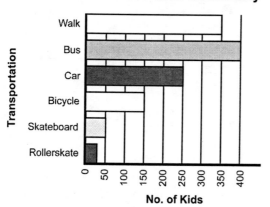

A. 200

B. 250

C. 150

D. 225

20. Solve.

$$\frac{7}{8} - \frac{3}{8} =$$

A. $\dfrac{10}{8}$

B. $\dfrac{3}{8}$

C. $\dfrac{8}{4}$

D. $\dfrac{4}{8}$

21. Which ordered pair represents the point on the coordinate plane above?

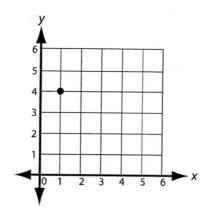

A. $(1, 4)$

B. $(4, 0)$

C. $(4, 1)$

D. $(1, 0)$

22. What is $4,546,713$ in word form?

A. Four billion, five hundred forty-six million, seven hundred thirteen

B. Four million, five hundred forty-six thousand, seven hundred thirteen

C. Four thousand five hundred forty-six thousand, seven hundred thirteen

D. Four million, five hundred sixty-four thousand, seven hundred thirteen

23. $458 \div 3 =$

A. 152 remainder 2

B. 153 remainder 1

C. 154 remainder 2

D. 152 remainder 3

24. Identify the correct operation.

$$12 \,\square\, 4 = 48$$

A. ÷

B. ×

C. −

D. +

M4A1b

25. Which is not a parallelogram?

A.

B.

C.

D.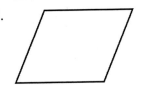

M4G1c

26. Round 147,543.2 to the nearest tens.

A. 147,544

B. 147,543

C. 147,542

D. 147,540

M4N2a

27. Which ordered pair represents the point on the coordinate plane above?

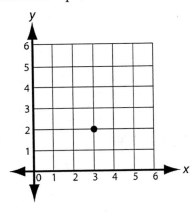

A. $(2, 3)$

B. $(2, 2)$

C. $(3, 2)$

D. $(3, 3)$

M4G3b

28. Put the following numbers in order from least to greatest.

8.7, 1.2, 5.6, 3.4

A. 1.2, 3.4, 8.7, 5.6

B. 8.7, 5.6, 3.4, 1.2

C. 1.2, 3.4, 5.6, 8.7

D. 1.2, 5.6, 3.4, 8.7

M4N5b

29. If Jeannie buys 3 pairs of socks at $1.99 for each pair and she buys a bag of pretzels for $2.59, about how much in change can Jeannie expect from a $10.00 bill to the nearest quarter?

A. $1.50

B. $1.75

C. $2.25

D. $3.00

M4N2d

5

30. Which number in the following sentence is the quotient?

$$8 \div 2 = 4$$

A. 4
B. 8
C. 2
D. None of the above

M4N4c

31. How many ounces are there in 1 pound?

A. 18
B. 16
C. 12
D. 8

M4M1c

32. Which of the following objects is not a rectangle?

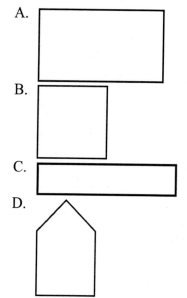

A.
B.
C.
D.

M4G1d

33. Given the numbers 3 and 9, choose the best description of their relationship.

A. 9 is twice as big as 3.
B. 3 is one half as large as 9.
C. 9 is three times larger than 3.
D. 3 is twice as large as 9.

M4A1a

34. Solve the problem below.

$$15.9 \div 3 =$$

A. 5.3
B. 3.5
C. 53
D. 35

M4N5e

35. Solve the problem below.

$$3\frac{2}{4} + 4\frac{1}{4} =$$

A. $1\frac{3}{4}$

B. $5\frac{3}{4}$

C. $7\frac{3}{4}$

D. $6\frac{3}{4}$

M4N6b

Part Two

1. Which digit is the ten-thousands place for 7,943,267?

   A. 4
   B. 9
   C. 3
   D. 7

   M4N1a

2. $4\frac{2}{4} + 6\frac{3}{4} =$

   A. 11
   B. $12\frac{1}{4}$
   C. $11\frac{1}{4}$
   D. 12

   M4N6b

3. The drawing of the skateboarder on a half pipe below shows what degree of rotation? (He starts and stops at the same point)

   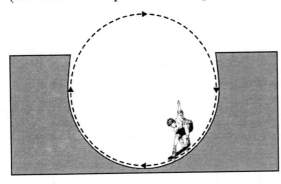

   A. 180°
   B. 120°
   C. 240°
   D. 360°

   M4M2b

4. Identify the correct operation.

   $16 \square 3 = 48$

   A. ÷
   B. −
   C. +
   D. ×

   M4A1b

5. Which ordered pair represents the point on the coordinate plane below?

   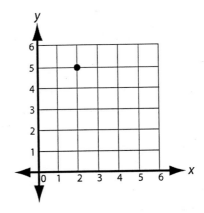

   A. $(5, 2)$
   B. $(2, 5)$
   C. $(2, 0)$
   D. $(0, 5)$

   M4G3b

6. Round 34,321 to the nearest hundred.

   A. 34,300
   B. 34,000
   C. 34,320
   D. 33,000

   M4N2a

7. Solve the following problem.

   $65 \div 9 =$

   A. 6 remainder 1
   B. 7 remainder 1
   C. 7 remainder 2
   D. 6 remainder 2

   M4N4b

8. Put the following numbers in order from largest to smallest.

　　3.4,　8.9　9.9　5.2　7.8

A. 8.9, 7.8, 9.9, 5.2, 3.4

B. 9.9, 8.9, 5.2, 7.8, 3.4

C. 9.9, 8.9, 7.8, 5.2, 3.4

D. 3.4, 5.2, 7.8, 8.9, 9.9

M4N5b

9. Using the graph below, how many more students preferred a praying mantis over crickets?

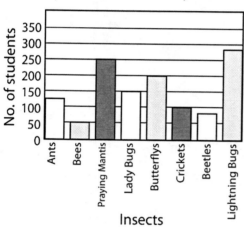

Favorite Bugs of 4th Graders
at I. Will Ketchum Elementary School

A. 125
B. 150
C. 350
D. 375

M4D1c

10. Round 45,786.5 to the nearest whole number.

A. 45,785

B. 45,780

C. 45,786

D. 45,787

M4N2c

11. Solve the following.

$$3 \times (2 + 4) - 3 \times (1 + 4) + 2 =$$

A. 5

B. 6

C. 4

D. 25

M4N7b

12. Jerry weighed exactly 7 pounds at birth. At the age of two, Jerry weighed 35 pounds. Jerry weighed 65 pounds at age nine and 130 pounds at age fifteen. How much weight did Jerry gain between the ages of two and nine?

A. 95

B. 28

C. 65

D. 30

M4M1c

13. Which of the triangles below is a right triangle?

A.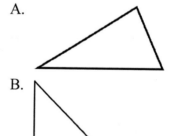

B.

C.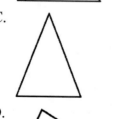

D.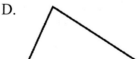

M4G1a

14. Solve.

$$550.8 \div 6 =$$

A. 93.8

B. 92.8

C. 91.8

D. 90.8

15. How many pounds are there in 4 tons?

A. 2,000

B. 4,000

C. 8,000

D. 5,000

16. Which pair of lines below are parallel?

A.

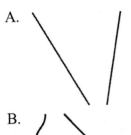

B.

C.

D.

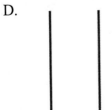

17. Multiply: 872×23.

A. 20,056

B. 20,156

C. 21,056

D. 20,055

18. Which design below contains a square?

A.

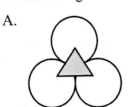

B.

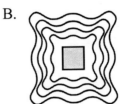

C.

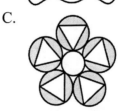

D.

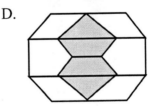

19. Write the number 34,562 in word form.

A. Thirty-four thousand, five hundred sixty-two

B. Thirty-four million, five hundred sixty-two thousand

C. Thirty-five thousand, four hundred sixty-two

D. Thirty-four thousand six hundred fifty-two

20. Which one answer is the correct one for both problems below?

$3{,}360 \div 30 =$
$336 \div 3 =$

A. 11
B. 110
C. 112
D. 111

M4N4d

21. Which fraction below is equivalent to $\frac{1}{2}$?

A. $\frac{3}{6}$

B. $\frac{4}{6}$

C. $\frac{3}{8}$

D. $\frac{2}{5}$

M4N6a

22. What is the measure of the angle?

A. 150°
B. 30°
C. 65°
D. 100°

M4M2a

23. Identify the correct operation.

$81 \,\triangle\, 9 = 9$

A. ×
B. ÷
C. −
D. +

M4A1b

24. Which figure below contains perpendicular lines, similar to a rectangular prism?

A.

B.

C.

D.

M4G2b

25. How many more students in the graph below prefer to play in the park instead of staying inside playing video games?

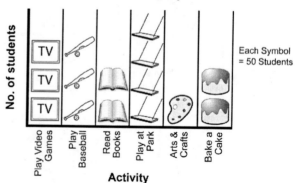

A. 50
B. 25
C. 75
D. 40

M4D1a

26. What place value is the digit 8 in the following number?

1,458,796

A. hundreds

B. thousands

C. tens

D. ten thousands

M4N1a

27. Amanda bought 6 pounds of potatoes at $1.00 per pound and 3 pounds of carrots at $0.75 per pound. To the nearest dollar, how much did Amanda spend on vegetables?

A. $7.00

B. $6.00

C. $9.00

D. $8.00

M4N2d

28. Which ordered pair represents the point on the coordinate plane below?

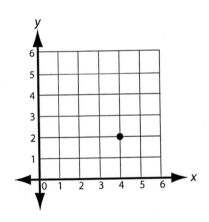

A. $(4, 2)$

B. $(2, 4)$

C. $(0, 4)$

D. $(4, 0)$

M4G3b

29. John Abram bought $1\frac{1}{2}$ pounds of hamburger at $3.00 per pound. What was the total cost of the hamburger?

A. $5.50

B. $4.50

C. $5.00

D. $4.00

M4N5c

30. Round $45,781.3$ to the nearest whole number.

A. 45,780

B. 45,782

C. 45,781

D. 45,779

M4N2c

31. Which of the following is a trapezoid?

A.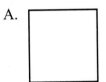

B.

C.

D.

M4G1c

Use the graph below to answer questions 32 and 33.

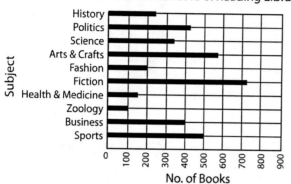

Subjects of Books Checked Out
in One Month at R. U. Reading Library

No. of Books

32. Which subject is checked out of the library the most in one month?

   A. Arts and Crafts

   B. Fiction

   C. Zoology

   D. History

   M4D1c

33. How many more sports books were checked out than fashion books?

   A. 200

   B. 300

   C. 250

   D. 350

   M4D1c

34. Solve.

$$3\frac{4}{8} + 2\frac{3}{8} =$$

   A. $6\frac{1}{8}$

   B. $5\frac{3}{8}$

   C. $5\frac{7}{8}$

   D. $6\frac{5}{8}$

   M4N6b

35. Add the following.

$$65.48 + 78.56 =$$

   A. 65.56

   B. 114.04

   C. 144.04

   D. 78.48

   M4N5c

Evaluation Chart for the Diagnostic Mathematics Test

Directions: On the following chart, circle the question numbers that you answered incorrectly. Then turn to the appropriate topics, read the explanations, and complete the exercises. Review the other chapters as needed. Finally, complete the *Passing the Georgia 4th Grade CRCT in Mathematics* Practice Tests to further review.

		Questions - Part 1	**Questions - Part 2**	**Pages**
Chapter 1:	Whole Numbers: Place Values and Rounding	2, 4, 10, 22, 29	1, 6, 19, 26, 27	15–26
Chapter 2:	Multiplication and Division	11, 15, 23, 30	7, 17, 20	27–34
Chapter 3:	Fractions	20, 35	2, 21, 34	35–45
Chapter 4:	Decimals	5, 14, 26, 28, 34	8, 10, 14, 29, 30, 35	46–58
Chapter 5:	Arithmetic Operations	12	11	59–68
Chapter 6:	Mathematical Relationships in Algebra	18, 24, 33	4, 23	69–78
Chapter 7:	Measurement	31	12, 15	79–85
Chapter 8:	Angles and Rotation	1, 3, 16	3, 22	86–94
Chapter 9:	Geometry	6, 9, 17, 25, 32	13, 16, 18, 31	95–103
Chapter 10:	Solid Geometry		24	104–110
Chapter 11:	Coordinate Systems	7, 21, 27	5, 28	111–117
Chapter 12:	Data Analysis	8, 13, 19	9, 25, 32, 33	118–132

NOTES

Chapter 1
Whole Numbers: Place Values and Rounding

This chapter covers the following Georgia Performance Standards:

M4N	Numbers and Operations	M4N1.a, b
M4N	Numbers and Operations	M4N2.a, b, d

1.1   Place Value: Greater Than One

Place Value: the value of a digit based upon its place within the number. For example, in the chart below, the digit 6 has two very different values in the number 6,271,863.

Millions	Hundred-thousands	Ten-thousands	Thousands	Hundreds	Tens	Ones
1,000,000	100,000	10,000	1,000	100	10	1
6	2	7	1	8	6	3

From the right, let's start in the ones column. As we progress from right to left, we see that each column represents 10 times the value of the column before it:

10 times 1 = 10

10 times 10 = 100

10 times 100 = 1,000

10 times 1,000 = 10,000

10 times 10,000 = 100,000

10 times 100,000 = 1,000,000

Using the place value chart on the previous page:

From right to left, the first six, listed in the tens column represents 6 tens.

6 times 10 = 60

Continuing to the second six from the right, we see a six with a very different value:

6 times 1,000,000 = 6,000,000.

Numbers can also be written in different forms. Notice the different values of the six when written in different forms.

The number written in **standard** form: 6,271,863

The number written in **expanded** form: 6,000,000 + 200,000 + 70,000 + 1,000 + 800 + 60 + 3 = 6,271,863.

The number written in **word** form reads as: six million, two hundred seventy-one thousand, eight hundred sixty-three.

Note: Be certain to put a comma in where necessary. A comma is placed every three digits, starting from the right.

Write the value of the number 2. The first one has been done for you.

1.   12,839   Two thousand (2,000)
2.   241,563
3.   12
4.   2,156,000

5.   28
6.   14,287
7.   628,566
8.   56,255

Use the number below to answer the following questions.

4,568,321

What number is in the –

9.   Tens place?
10. Millions place?
11. Hundred-thousands place?
12. Ones place?

13. Hundreds place?
14. Thousands place?
15. Ten-thousands place?

Write the value of the underlined digit. The first one has been done for you.

16.   47<u>5</u>   five ones
17.   <u>3</u>,889,778
18.   5<u>6</u>7
19.   <u>2</u>4,221

20.   <u>6</u>,112,345
21.   <u>1</u>57,812
22.   <u>8</u>4
23.   4,4<u>2</u>9

1.2   Numbers in Expanded Notation

Let's practice writing numbers in expanded notation. **Expanded notation** is the number written out with each of its digits multiplied by its column.

Example 1:	Write the number 243 in expanded notation.
	The number 243 has 2 in the hundreds place, 4 in the tens place, and 3 in the ones place.
	This means there are 2 groups of hundreds, 4 groups of tens, and 3 ones.
Answer:	Expanded notation of 243 is $2 \times 100 + 4 \times 10 + 3 \times 1$.
Example 2:	Write 4,182,637 in expanded notation.
	Find the place value of each digit. The number 4,182,637 has 4 in the millions place, 1 in the hundred-thousands place, 8 in the ten-thousands place, 2 in the thousands place, 6 in the hundreds place, 3 in the tens place, and 7 in the ones place.
Answer:	Expanded notation is $4 \times 1,000,000 + 1 \times 100,000 + 8 \times 10,000 + 2 \times 1,000 + 6 \times 100 + 3 \times 10 + 7 \times 1$.

Write the following in expanded notation.

1. 243

2. 6, 458

3. 1,446,879

4. 621,378

5. 52

6. 36,112

7. 174,224

8. 862

9. 47,883

10. 777

11. 24,600

12. 9,118,766

1.3   Writing Numbers in Word Form

Being able to read and write whole numbers is very important. For instance, someday you may open up a checking account. You will need to be able to write the amount of the check in both standard form and word form. If a bank teller cannot read the amount written in standard form, it is the law that the teller must deposit the check according to the word form on the check. This also applies to writing a check at a store. If the person writing the check makes a mistake and writes a different amount in the standard form from the word form, the cashier, by law, must use the word form.

Steven Jones
Macon, GA

8/4/2012

Pay to the order of: Winslow Abrams   Amount: $ 3,200.00

Three thousand, two hundred dollars and 00/100 ******   dollars

Steven Jones

Imagine you wrote $494.00 in the standard form place on the check, but mistakenly wrote in the word form place on the check, four hundred ninety nine dollars and $\frac{00}{100}$. Your check will be cashed for five dollars more than you intended! You can see, it is very important to know how to write the word form of a number! (Note: The cents on the check are often written $\frac{cents}{1\ dollar}$.)

Jimmy Smith
Winder, GA

8/4/2012

Pay to the order of: Dr. Luke Rockland   Amount: $ 494.00

Four hundred ninety nine and 00/100*****   dollars

Jimmy Smith

When writing numbers in word form, the numbers, like any sentence, are read from left to right. Starting on the **right**, every 3 numbers are separated by a comma, just like they are in standard form. Use the chart below as a model of the placement value of each number. Notice where the commas go.

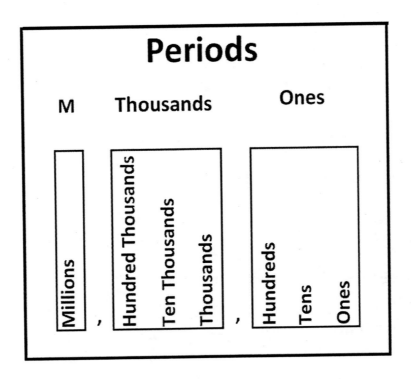

Example 3:   **Write the number 78,624 in word form.**

Step 1:   Write out the first set of numbers as is: seventy-eight. Because it is the thousands place, you will add the word "thousand" to the end of the first set of numbers: seventy-eight thousand.

Step 2:   Now do the next set of numbers: six hundred twenty-four. Just combine the two sets of numbers with a comma in between them.

Solution: Seventy-eight thousand, six hundred twenty-four.

Write the following numbers in word form.

1. 451

2. 2,112,222

3. 6,547

4. 27

5. 812,321

6. 56,712

7. 367

8. 5,444

9. 4,887,999

10. 19,125

11. 872

12. 12

13. 3,333

14. 7,134,628

15. 288

In the following exercise, write the numbers in standard form.

16. eight hundred twenty-two

17. one million, three hundred thousand, two hundred twelve

18. forty-seven thousand, nine hundred two

19. thirteen

20. seven hundred sixty-three thousand, nine hundred seventeen

21. forty-two

22. eight hundred ninety-nine

23. seven million, six hundred eighty-three thousand, three hundred sixty-eight

24. seven hundred sixteen

1.4 Rounding Whole Numbers

Millions	**Hundred Thousands**	**Ten Thousands**	**Thousands**	**Hundreds**	**Tens**	**Ones**
7,	8	2	6,	4	5	9

Example:

Consider the number 7,826,459 shown above with the place values labeled. To round to a given place value, first find the digit in that place value. Then look to the digit on the right. If the digit on the right is 5 or greater, INCREASE the given place value BY ONE. All the digits to the right of the given place value become 0's. If the digit on the right is LESS THAN 5, leave the place value digit the same, and just change the digits on the right to 0's.

Where might you use your skill at rounding numbers? You might round numbers whenever you want to quickly estimate an amount of something without having to add it up exactly. For instance: suppose you have $10.00 to go shopping at the mall. You decide to spend it all at the toy store, and you want to be sure you have enough money for everything you want. Your kite costs $1.89, your jump rope costs $1.29, and your joke book costs $5.99. If you round the kite to $2.00 (up from $1.89), round the jump rope to $1.00 (down from $1.29), and round the joke book to $6.00 (up from $5.99), you can see quickly that you will need $2.00 + $1.00 + $6.00 = $9.00. That's much faster to estimate the total cost than adding it up exactly.

Another example where rounding numbers comes in handy is talking about very large numbers. For instance, if the Governor wants to know how much water is used in one day in Georgia, he or she will want the number of MILLIONS of gallons used. He or she doesn't really need the number down to the single gallon.

Or suppose you want to know how many candies are in a bag with different size pieces without counting the whole bag. Say you weigh out one ounce on a scale and you count 9 pieces. Then you weigh out another ounce on a scale and count 11 pieces. You can round to the nearest ten and find there are about 10 pieces of candies for each ounce. If you know how many ounces are in the bag, you simply multiply the number of ounces by ten. This should give you a close estimate of how many pieces of candy are in the bag.

Example 4:   Round the number 7,826,459 to the nearest:

ten	7,826,460
hundred	7,826,500
thousand	7,826,000
ten thousand	7,830,000
hundred thousand	7,800,000
million	8,000,000

Round to the nearest ten.

1.   678

2.   8,725

3.   5,672,813

4.   27,855

Round to the nearest hundred.

5.   751

6.   2,645,289

7.   4,720

8.   13,372

Round to the nearest thousand.

9.   7,821

10.   12,853

11.   3,221

12.   8,741,501

Round to the nearest ten thousand.

13.   6,451,224

14.   4,221,555

15.   45,102

16.   112,999

Round to the nearest hundred thousand.

17.   651,471

18.   1,251,100

19.   964,968

20.   4,732,332

Round to the nearest million.

21.   2,361,241

22.   7,541,223

23.   1,142,999

24.   5,654,921

Figure the following problems using your new skill of rounding numbers.

25. To the nearest thousand dollars, how much money will Mrs. Robertson need to buy 4 new TVs for the school if each TV costs $499.00?

26. To the nearest ten pounds, how many pounds do Mula's brothers and sisters weigh? Carrie weighs 52 pounds, Steven weighs 18 pounds, Oscar weighs 48 pounds, and Gretchen weighs 73 pounds.

27. The mayor would like to estimate the traffic congestion of the four new subdivisions in his city: Applegate subdivision = 120 homes, Dogwood Breeze subdivision = 432 homes, Peach Estates subdivision = 750 homes, and Green Mountain subdivision = 342 homes. To the nearest hundred, how many homes are in the four subdivisions?

28. To the nearest ten thousand, estimate the number of eggs that the grocery store chain will buy from the following three farmers: Farmer Brown has 52,000 eggs, Farmer Kimson has 87,000, and Farmer Huckleberry has 64,000.

29. The school librarian wants to estimate how many books are checked out in a week. On Monday 94 books were checked out, 62 books were checked out on Tuesday , Wednesday 112 books, Thursday 157 books, and Friday 55 books. How many books were checked out for the week to the nearest hundred?

30. Ms. Samuels has two 20 dollar bills to buy markers for her class. If she has 18 students, and each packet of markers costs $1.75, will she have enough money to get each of her students markers?

Chapter 1 Review

Write the value of the digit 8 in the next three problems.

1. 28

2. 4,845,224

3. 78,667

Write the following numbers in expanded form.

4. 482

5. 14,623

6. 9,887

Write the following numbers in word form.

7. 225

8. 6,450

9. 12

10. 10,549

11. 122,729

12. 2,340,063

13. 32,879

14. 583

Round the following numbers.

15. Round to the nearest ten: 657

16. Round to the nearest thousand: 34,299

17. Round to the nearest hundred: 7,324,890

18. Round to the nearest ten thousand: 877,347

Answer the following.

19. Lisa has three ten dollar bills in her purse. If she buys 3 packages of socks at $4.50 each, how many ten dollar bills should Lisa give to the cashier?

20. Mr. Handingson is planting lettuce seeds for a large garden. There are approximately 115 seeds per package. If Mr. Handingson wants to plant 300 lettuce plants, how many packages will he need to buy?

21. Mrs. Trent, the lunch room lady, has to order pizzas for the school. Each large pizza has ten pieces and there are 2,456 students in the school. The pizza manufacturer sells the pizzas in boxes of 100. How many boxes will Mrs. Trent need to order for each student to get one slice?

Chapter 1 Test

1. What is the place value of the underlined digit?

    12,662

    A. hundreds

    B. thousands

    C. tens

    D. ten-thousands

2. What is $6 \times 1,000,000 + 8 \times 1,000 + 6 \times 100 + 3 \times 10$ in standard form?

    A. 681,631
    B. 6,008,630
    C. 6,181,613
    D. 608,630

3. What is 65,982 in word form?

    A. sixty-five hundred, nine hundred eighty-two

    B. sixty-five thousand, eight hundred ninety-two

    C. sixty-five thousand, nine hundred eighty-two

    D. sixty-five hundred, ninety-two

4. Round 4,578,990 to the nearest hundred thousand.

    A. 4,600,000

    B. 5,000,000

    C. 4,580,000

    D. 4,578,000

5. Round the total weight of boxes of toys to be shipped to the nearest ten: 41 pounds, 62 pounds and 78 pounds.

    A. 190

    B. 170

    C. 160

    D. 180

6. What is the place value of the five in 7,050,000?

    A. hundreds

    B. thousands

    C. millions

    D. ten-thousands

7. What is eight hundred forty-two thousand, six hundred sixty-three in standard form?

    A. 842,663

    B. 84,263

    C. 842,633

    D. 84,633

8. Round 555,555 to the nearest hundred.

    A. 555,600

    B. 556,000

    C. 556,600

    D. 555,660

9. Hank has 7 five dollar bills in his pocket. He needs to buy soft drinks for a family picnic that cost $4.95 and disposable cups at $2.99 and paper plates at $2.49. How many five dollar bills should Hank give the cashier to pay for his items?

A. 2

B. 3

C. 4

D. 5

10. What is 222 in expanded form?

A. two hundred twenty two

B. $200 + 20 + 2$

C. two hundred thousand

D. $2 \times 1,000 + 2 \times 10 + 24 \times 1$

11. What is 711,800 in word form?

A. seven hundred thousand, eight hundred

B. seven hundred eleven thousand, eight hundred

C. seven hundred eleven thousand, eighty

D. seventy-one thousand, eight hundred

12. Round 87,999 to the nearest thousand.

A. 87,900

B. 88,000

C. 90,000

D. 88,900

STOP! You are done with the test!

Chapter 2
Multiplication and Division

This chapter covers the following Georgia Performance Standards:

| M4N | Numbers and Operations | M4N3 |
| M4N | Numbers and Operations | M4N4.a, b, c, d |

2.1 Multiplying Whole Numbers

Example 1: Multiply: 256×73

Step 1: Line up the ones digits. Multiply 256×3.

$$
\begin{array}{r}
\overset{1\ 1}{2\,5\,6} \\
\times\ 7\boxed{3} \\
\hline
7\,6\,8
\end{array}
$$

$3 \times 6 = 18$, write 8 and carry the one
$3 \times 5 = 15$, add the 1 that was carried to get 16, write 6 and carry the one
$3 \times 2 = 6$, add the 1 that was carried to get 7, write 7

Step 2: Multiply 256×7. Remember to shift the product one place to the left. Then add.

$$
\begin{array}{r}
\overset{3\ 4}{2\,5\,6} \\
\times\ \boxed{7}3 \\
\hline
7\,6\,8 \\
1\,7\,9\,2\ \ \\
\hline
1\,8,6\,8\,8
\end{array}
$$

$7 \times 6 = 42$, write 2 and carry the 4
$7 \times 5 = 35$, add the 4 that was carried to get 39, write 9 and carry the 3
$7 \times 2 = 14$, add the 3 that was carried to get 17, write 17

└── Add

Answer: 18,688

Multiply.

1. $\begin{array}{r} 246 \\ \times\ \ 24 \\ \hline \end{array}$

2. $\begin{array}{r} 683 \\ \times\ \ 11 \\ \hline \end{array}$

3. $\begin{array}{r} 87 \\ \times\ \ 32 \\ \hline \end{array}$

4. $\begin{array}{r} 999 \\ \times\ \ 12 \\ \hline \end{array}$

5. $\begin{array}{r} 112 \\ \times\ \ 12 \\ \hline \end{array}$

6. $\begin{array}{r} 54 \\ \times\ \ 19 \\ \hline \end{array}$

7. $\begin{array}{r} 16 \\ \times\ \ 23 \\ \hline \end{array}$

8. $\begin{array}{r} 884 \\ \times\ \ 48 \\ \hline \end{array}$

9. $\begin{array}{r} 142 \\ \times\ \ 95 \\ \hline \end{array}$

10. $\begin{array}{r} 164 \\ \times\ \ 4 \\ \hline \end{array}$

11. $\begin{array}{r} 342 \\ \times\ \ 25 \\ \hline \end{array}$

12. $\begin{array}{r} 83 \\ \times\ \ 24 \\ \hline \end{array}$

13. $\begin{array}{r} 781 \\ \times\ \ 13 \\ \hline \end{array}$

14. $\begin{array}{r} 345 \\ \times\ \ 12 \\ \hline \end{array}$

15. $\begin{array}{r} 49 \\ \times\ \ 7 \\ \hline \end{array}$

16. $\begin{array}{r} 68 \\ \times\ \ 42 \\ \hline \end{array}$

17. $\begin{array}{r} 111 \\ \times\ \ 66 \\ \hline \end{array}$

18. $\begin{array}{r} 44 \\ \times\ \ 27 \\ \hline \end{array}$

19. $\begin{array}{r} 118 \\ \times\ \ 68 \\ \hline \end{array}$

20. $\begin{array}{r} 129 \\ \times\ \ 23 \\ \hline \end{array}$

2.2 Dividing Whole Numbers

Example 2:   Divide: $4{,}993 \div 24$

Step 1:   Rewrite the problem using the symbol $\overline{)}$.

Step 2:   Divide 49 by 24. Multiply 2×24, and subtract.

$$
\begin{array}{r}
2 \\
24\overline{)4993} \\
-48\downarrow \\
\hline
19
\end{array}
$$

Step 3:   You will notice you cannot divide 19 by 24. You must put a 0 in the answer and then bring down the 3.

$$
\begin{array}{r}
20 \\
24\overline{)4993} \\
-48\downarrow \\
\hline
193
\end{array}
$$

Step 4:   Divide 193 by 24. Multiply 8×24, and subtract.

$$
\begin{array}{r}
208\,r\,1 \\
24\overline{)4993} \\
-48 \\
\hline
193 \\
-192 \\
\hline
1
\end{array}
$$

The answer is 208 with a remainder of 1.

Divide.

1. $864 \div 24$ 3. $452 \div 18$ 5. $999 \div 11$ 7. $633 \div 3$

2. $6{,}040 \div 40$ 4. $456 \div 19$ 6. $1{,}071 \div 18$ 8. $325 \div 18$

9. $7,156 \div 65$ 12. $4,520 \div 25$ 15. $6,000 \div 30$ 18. $378 \div 9$

10. $100 \div 11$ 13. $1,240 \div 40$ 16. $857 \div 55$ 19. $176 \div 44$

11. $1,200 \div 51$ 14. $444 \div 11$ 17. $296 \div 37$ 20. $432 \div 12$

2.3   The Parts of a Division Problem

Let's pick apart a division problem and learn some vocabulary words! Each division problem has three parts: a **dividend,** a **divisor,** and a **quotient**.

Dividend ÷ Divisor = Quotient

As we saw in the problems in part one of this chapter, sometimes the quotient includes a remainder– the part of the answer that is smaller than the divisor and is a "leftover", so to speak.

In the following problem, let's identify the parts: $240 \div 25 = 9 \text{ r}15$.
240 is the dividend, 25 is the divisor, 9 r15 is the quotient.

Name the parts as requested below.

1. What is the dividend in $81 \div 9 = 9$?

2. What is the divisor in $100 \div 20 = 5$?

3. What is the quotient in $50 \div 3 = 16 \text{ r}2$?

4. What is the divisor in $17 \div 2 = 8 \text{ r}1$?

5. What is the dividend in $682 \div 47 = 14 \text{ r}24$?

6. What is the quotient in $18 \div 3 = 6$?

7. What is the divisor in $1,000 \div 100 = 10$?

8. What is the dividend in $888 \div 11 = 80 \text{ r}8$?

Dividing by a multiple of 10 is easy. The example below will show you how to make division problems that have multiples of ten easier to solve.

Example 3:   Divide $2,700 \div 100$

   Step 1:   Count the number of 0's in the multiple of 10. There are two 0's in 100.

   Step 2:   Take two zeros away from the dividend in the problem and write the answer. $2,700 \div 100 = 27$. You can see at a glance that both $2,700 \div 100 = 27$ and $27 \div 1 = 27$. Both problems have the same answer.

Find the quotient for each of the two problems below.

9. $7,000 \div 1,000$ and $7 \div 1$

10. $800,000 \div 100$ and $8,000 \div 1$

11. $200,000 \div 1,000$ and $200 \div 1$

12. $2,050 \div 50$ and $205 \div 5$

13. $60,800 \div 200$ and $608 \div 2$

14. $745,000 \div 250$ and $74,500 \div 25$

15. $444,000 \div 1,000$ and $444 \div 1$

16. $650 \div 50$ and $65 \div 5$

17. $1,000,000 \div 1,000$ and $1,000 \div 1$

18. $804,000 \div 400$ and $8,040 \div 4$

19. $1,670,000 \div 5,000$ and $1,670 \div 5$

20. $10,000 \div 250$ and $1,000 \div 25$

21. $4,200,000 \div 2,100,000$ and $42 \div 21$

22. $999,000 \div 300$ and $9,990 \div 3$

Across the Great Divide and on to the CHAPTER Review

Chapter 2 Review

Solve the following multiplication problems.

1. 612×12

2. 893×25

3. 112×57

4. 921×15

5. 65×27

6. 44×22

7. 821×43

8. 651×11

Solve the following division problems.

9. $258 \div 12$

10. $640 \div 8$

11. $8,020 \div 20$

12. $6,100,100 \div 25$

13. $842 \div 12$

14. $627 \div 37$

15. $6,453 \div 78$

16. $2,111 \div 11$

Answer the following questions.

17. Which number is the dividend in $456 \div 21 = 21 \text{ r}15$?

18. Which number is the divisor in $14 \div 7 = 2$?

19. Which number is the remainder in $29 \div 9 = 3 \text{ r}2$?

20. Which number is the quotient in $30 \div 6 = 5$?

Solve the following using your new skill of dividing by multiples of ten.

21. $880,000 \div 88$

22. $7,410 \div 10$

23. $5,647,000 \div 500$

24. $600 \div 60$

25. $4,050 \div 50$

26. $3,600,000 \div 600$

27. $80,000 \div 4,000$

28. $100,000 \div 1,000$

29. $460 \div 230$

30. $3,174,000 \div 2,000$

Chapter 2 Test

1. What is the dividend in $400 \div 2 = 200$?

   A. 400

   B. 2

   C. 200

   D. $\div$

2. Divide: $6,500,000 \div 65$

   A. 6,500

   B. 100,000

   C. 650

   D. 10,000

3. Multiply: 248×12

   A. 2,480

   B. 2,796

   C. 24,800

   D. 2,976

4. Divide: $4,521 \div 15$

   A. 301 r6

   B. 301

   C. 301 r8

   D. 301 r4

5. Which number is the quotient in $99 \div 11 = 9$?

   A. 99

   B. $\div$

   C. 9

   D. 11

6. Divide: $400,000 \div 4,000$

   A. 100

   B. 1,000

   C. 10

   D. 10,000

7. Multiply: 622×34

   A. 21,140

   B. 21,142

   C. 21,148

   D. 12,148

8. Which number is the divisor in $12 \div 3 = 4$?

   A. 12

   B. 3

   C. 4

   D. $\div$

9. Divide: $8,000,100 \div 100$

   A. 8,000

   B. 8,001

   C. 80,001

   D. 80,000

10. Multiply: 742×88

    A. 56,290

    B. 6,529

    C. 65,269

    D. 65,296

11. Divide: $3,871 \div 12$

    A. 322 r7

    B. 322 r11

    C. 322 r3

    D. 321 r7

12. Divide: $20,000 \div 500$

    A. 400

    B. 40

    C. 4,000

    D. 4

Practice and Fun!

Do the problems in the clues and fill in the cross-number puzzle below.

Across

1. 111×3
3. $150 \div 2$
4. 9×7
6. 37×25
8. 164×4
10. 87×5

11. 12×6
13. $144 \div 2$
14. 14×9
15. 423×170
16. $936 \div 3$
17. 9×11

19. $1,000 \div 100$
20. $72 \div 2$
22. 3×33
23. 13×6
24. $936 \div 39$
25. $1,575 \div 3$

27. 163×5
29. 4×16
30. $88 \div 2$
32. $98,300 \div 100$
33. 12×12

Down

2. 61×5
3. $7,222,970 \div 10$
5. $76,942 \times 5$
7. 71×3

9. $36 \div 6$
12. 313×3
18. 116×8
19. $133 \div 7$

21. 8×8
22. $51,072 \div 56$
26. 81×3
28. 11×11

29. $33,610 \div 5$
31. 157×31
32. $16,812 \div 18$
34. 61×8

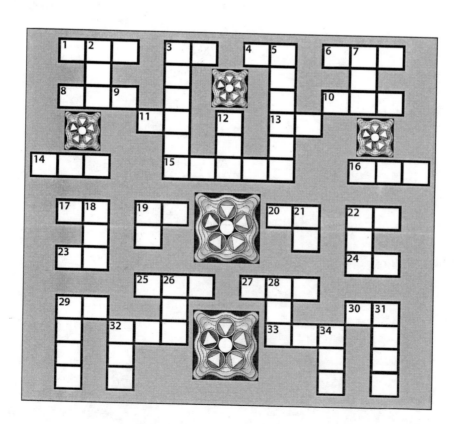

Chapter 3
Fractions

This chapter covers the following Georgia Performance Standards:

| M4N | Numbers and Operations | M4N6.a, b, c |

3.1   Fraction Equivalents

Fractions are numbers that represent how many parts of a whole number there are. An example of a fraction is $\frac{3}{4}$. The top part of the fraction, 3, is called the **numerator**, which tells how many parts are included. The bottom part of the fraction, 4, is the **denominator**, which tells how many parts in total there are. In the case of $\frac{3}{4}$, the fraction is telling you that 3 out of 4 parts are included.

$$\frac{3}{4} = \frac{\text{Numerator}}{\text{Denominator}}$$

Example 1:   A blueberry cobbler is cut into 12 pieces. 9 of the pieces have been eaten by the Hanson family. How much of the cobbler has been eaten?

Step 1:   Draw a pan of cobbler and divided the pan into 12 equal parts. Shade the 9 parts that have been eaten.

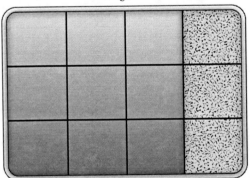

Blueberry Cobbler

Step 2:   Write the fraction of the parts that have been eaten.

$$\frac{\text{how many pieces were eaten}}{\text{total number of pieces}} = \frac{9}{12}$$

Answer: $\frac{9}{12}$ of the pan of blueberry cobbler was eaten, which is equivalent to $\frac{3}{4}$ of the entire cobbler.

Write a fraction for the part of the figure that is shaded.

1.
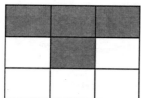

3.
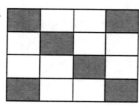

5.
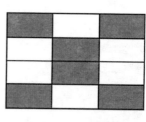

6.
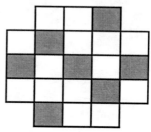

2.
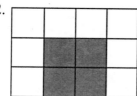

4.
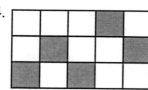

3.2   Adding Fractions and Mixed Numbers

Adding fractions is similar to adding whole numbers when the fractions have the same denominator.

Example 2:   **Add the fractions.** $\frac{3}{11} + \frac{2}{11}$

Step 1:   Both fractions have a common denominator, 11. To add two fractions that share a common denominator, simply add the two numerators together. $3 + 2 = 5$.

Step 2:   Put the total of the two numerators over the common denominator of the fraction. $\frac{5}{11}$

Answer: $\frac{5}{11}$

Example 3:   **Add the mixed numbers.** $3\frac{2}{8} + 2\frac{4}{8}$

Step 1:   Add the fractions first, starting with the numerator. $2 + 4 = 6$

Step 2:   Place the total of the two numerators, 6, over the common denominator of the fraction. $\frac{6}{8}$

Step 3:   Add the whole numbers. $3 + 2 = 5$

Step 4:   Place the sums of the fraction and whole number together. $5\frac{6}{8}$

Answer: $5\frac{6}{8}$

Add the fractions.

1. $\frac{2}{5} + \frac{2}{5}$

2. $\frac{3}{12} + \frac{7}{12}$

3. $\frac{1}{9} + \frac{7}{9}$

4. $\frac{3}{5} + \frac{2}{5}$

5. $1\frac{6}{10} + 6\frac{1}{10}$

6. $\frac{1}{7} + \frac{3}{7}$

7. $4\frac{8}{11} + 3\frac{2}{11}$

8. $\frac{2}{4} + \frac{1}{4}$

9. $\frac{4}{12} + \frac{5}{12}$

10. $\frac{1}{6} + \frac{3}{6}$

11. $\frac{1}{5} + \frac{2}{5}$

12. $\frac{7}{11} + \frac{2}{11}$

13. $2\frac{3}{7} + 9\frac{3}{7}$

14. $\frac{2}{10} + \frac{7}{10}$

15. $\frac{1}{7} + \frac{2}{7}$

Adding fractions in word problems.

16. Aunt Annie baked an apple pie. Her nephews ate $\frac{2}{6}$ of the pie and her nieces ate $\frac{3}{6}$ of the pie. How much pie was eaten?

17. Zach and Jake shared a quart of soda. Zach drank $\frac{1}{4}$ of the soda and Jake drank $\frac{2}{4}$ of the soda. What fraction of the soda did both Zach and Jake drink?

18. Jason and Enzo were playing marbles. There were 12 large blue glass marbles between them. Jason had $\frac{2}{12}$ of the large blue glass marbles and Enzo had $\frac{7}{12}$ of the large blue glass marbles. The rest of the large blue glass marbles were just knocked out of the game by Jason's cat named Trouble. What fraction of the large blue glass marbles did Jason and Enzo still have?

19. Mara and Irene were making 3 patchwork quilts together. Mara had sewn $1\frac{2}{8}$ of the quilts and Irene had sewn $\frac{3}{8}$ of the quilts. How much of the quilts have Mara and Irene completed?

20. Tai and Tyler planned on playing 10 games of ping-pong after school on Friday. Tai won $\frac{4}{10}$ of the ping-pong games and Tyler won $\frac{3}{10}$ of the ping-pong games so far. What fraction of the 10 games of ping-pong have Tai and Tyler played?

3.3   Subtracting Fractions and Mixed Numbers

Subtracting fractions is similar to subtracting whole numbers when the fractions have the same denominator.

Example 4:   **Subtract the fractions.** $\frac{9}{10} - \frac{2}{10}$

Step 1:   Both fractions have a common denominator, 10. To subtract two fractions that share a common denominator, simply subtract the two numerators. $9 - 2 = 7$.

Step 2:   Put the difference of the two numerators over the common denominator of the fraction. $\frac{7}{10}$

Answer: $\frac{7}{10}$

Example 5:   **Subtract the fractions.** $\frac{7}{11} - \frac{4}{11}$

Step 1:   Subtract the numerators of the fractions. $7 - 4 = 3$

Step 2:   Place the difference of the two numerators, 3, over the denominator of the fraction. $\frac{3}{11}$

Answer: $\frac{3}{11}$

Example 6:   **Subtract the mixed numbers.** $4\frac{6}{12} - 1\frac{5}{12}$

Step 1:   Since both fractions have a common denominator, you subtract the numerators. $6 - 5 = 1$

Step 2:   Place the difference of the two numerators, 1, over the common denominator of the fractions. $\frac{1}{12}$

Step 3:   Subtract the whole numbers. $4 - 1 = 3$

Step 4:   Place the differences of the fraction and whole number together. $3\frac{1}{12}$

Answer: $3\frac{1}{12}$

Subtract the fractions.

1. $\dfrac{8}{9} - \dfrac{3}{9}$

2. $\dfrac{11}{12} - \dfrac{4}{12}$

3. $\dfrac{7}{8} - \dfrac{2}{8}$

4. $5\dfrac{3}{7} - 2\dfrac{1}{7}$

5. $\dfrac{9}{10} - \dfrac{5}{10}$

6. $\dfrac{5}{6} - \dfrac{3}{6}$

7. $\dfrac{8}{11} - \dfrac{4}{11}$

8. $\dfrac{4}{5} - \dfrac{2}{5}$

9. $10\dfrac{4}{4} - 7\dfrac{3}{4}$

10. $\dfrac{2}{3} - \dfrac{1}{3}$

11. $6\dfrac{8}{12} - \dfrac{7}{12}$

12. $\dfrac{6}{7} - \dfrac{4}{7}$

13. $\dfrac{7}{9} - \dfrac{3}{9}$

14. $\dfrac{6}{8} - \dfrac{4}{8}$

15. $\dfrac{8}{10} - \dfrac{5}{10}$

Subtracting fractions in word problems.

16. Chalee was baking a pan of oatmeal cookie bars for a bake sale at Rose Valley Elementary. Sadly, Chalee's dog jumped up on the table and ate $\frac{1}{4}$ of the pan of oatmeal cookie bars while Chalee was doing her homework. What fraction of the pan of oatmeal cookie bars were left after Chalee's dog got to them? (Hint: the full pan of bars $= \frac{4}{4}$)

17. Isaias made a goal to spend $3\frac{3}{4}$ hours every week practicing his French horn for music class. Today, Isaias has already played $1\frac{2}{4}$ hours. How much more practice does Isaias need to do in order to meet his goal?

18. Shamindra walks her dog named Volley 6 hours every week. It's good exercise for both Shamindra and Volley! So far this week, Shamindra has walked Volley for 3 hours. What fraction of the 6 hours does Shamindra and Volley have left to walk?

19. Paul and Kevin are selling video games at a yard sale to raise money for the local animal shelter. Altogether, the boys decided to sell 10 games of their video game collection. The boys sold $\frac{7}{10}$ of the games. What fraction of the video games are left to sell?

20. Arnie had to shovel all the sand out of his backyard sandbox so he could replace the old sand with fresh sand. Arnie knew that he needed 9 bags in all to fill the sand box. By 2:00 in the afternoon, Arnie had shoveled $\frac{5}{9}$ of the sand out of the sand box. What fraction of the sand still needs to be shoveled out of the sand box?

3.4   Changing Mixed Numbers to Improper Fractions

A **mixed number** has both a whole number and a fraction together. An improper fraction is a fraction that has a larger numerator than denominator.

Example 7:   Change $2\frac{2}{3}$ to an improper fraction.

   Step 1:   Multiply the whole number, 2, by the denominator of the fraction, 3.

$$2 \times 3 = 6$$

   Step 2:   Add the numerator to the product from Step 1.

$$6 + 2 = 8$$

   Step 3:   Put the answer from Step 2 over the denominator, 3.

$$2\frac{2}{3} = \frac{8}{3}$$

Answer: $\frac{8}{3}$

Change the following mixed numbers to improper fractions.

1. $7\frac{2}{5} =$ _____

2. $3\frac{1}{4} =$ _____

3. $4\frac{4}{7} =$ _____

4. $5\frac{1}{5} =$ _____

5. $11\frac{2}{6} =$ _____

6. $7\frac{2}{3} =$ _____

7. $9\frac{1}{2} =$ _____

8. $4\frac{3}{6} =$ _____

9. $7\frac{3}{8} =$ _____

10. $6\frac{4}{5} =$ _____

11. $2\frac{2}{4} =$ _____

12. $7\frac{5}{7} =$ _____

13. $2\frac{3}{8} =$ _____

14. $10\frac{7}{9} =$ _____

15. $3\frac{1}{7} =$ _____

16. $4\frac{3}{4} =$ _____

17. $5\frac{2}{3} =$ _____

18. $11\frac{6}{7} =$ _____

19. $2\frac{3}{5} =$ _____

20. $12\frac{1}{3} =$ _____

Whole numbers become improper fractions when you put them over 1. Change the following whole numbers to improper fractions. The first one is done for you.

21. $6 = \frac{6}{1}$

22. $12 =$ ____

23. $5 =$ ____

24. $3 =$ ____

25. $10 =$ ____

26. $4 =$ ____

27. $1 =$ ____

28. $2 =$ ____

29. $7 =$ ____

30. $11 =$ ____

3.5   Changing Improper Fractions to Mixed Numbers

Improper fractions have a larger numerator than the denominator. Improper fractions do not usually stand as a final answer in a math problem unless the directions say to leave the answer as an improper fraction. In the last lesson, you learned how to change a mixed number into an improper fraction. Now let's turn an improper fraction back into a mixed number.

Example 8:   Change $\dfrac{14}{3}$ to a mixed number.

Step 1:   Divide the numerator, 14, by the denominator of the fraction, 3.

$$14 \div 3 = 4 \text{ r } 2$$

Step 2:   The whole number in the quotient from Step 1 becomes the whole number in the mixed number, 4.

Step 3:   The remainder, 2, becomes the numerator in the fractions. The denominator from the mixed number, 3, remains the same.

$$\frac{14}{3} = 4\frac{2}{3}$$

Answer: $4\dfrac{2}{3}$

Change the following improper fractions into mixed fractions.

1. $\dfrac{17}{8}$ 6. $\dfrac{34}{11}$ 11. $\dfrac{23}{6}$ 16. $\dfrac{14}{11}$

2. $\dfrac{23}{5}$ 7. $\dfrac{9}{4}$ 12. $\dfrac{7}{2}$ 17. $\dfrac{67}{8}$

3. $\dfrac{11}{2}$ 8. $\dfrac{29}{9}$ 13. $\dfrac{82}{9}$ 18. $\dfrac{17}{5}$

4. $\dfrac{13}{8}$ 9. $\dfrac{7}{3}$ 14. $\dfrac{38}{6}$ 19. $\dfrac{17}{3}$

5. $\dfrac{25}{6}$ 10. $\dfrac{15}{4}$ 15. $\dfrac{29}{7}$ 20. $\dfrac{26}{10}$

Chapter 3 Review

Write a fraction for the part of the figure that is shaded.

1.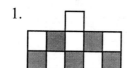

2.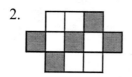

3.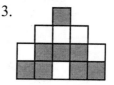

Add the fractions.

4. $\dfrac{4}{12} + \dfrac{7}{12} =$ _____

5. $\dfrac{3}{7} + \dfrac{2}{7} =$ _____

6. $3\dfrac{2}{6} + 3\dfrac{3}{6} =$ _____

7. $\dfrac{4}{12} + \dfrac{5}{12} =$ _____

8. $\dfrac{5}{11} + \dfrac{4}{11} =$ _____

9. $2\dfrac{1}{3} + 7\dfrac{1}{3} =$ _____

Subtract the fractions.

10. $\dfrac{11}{12} - \dfrac{5}{12} =$ _____

11. $\dfrac{9}{10} - \dfrac{6}{10} =$ _____

12. $4\dfrac{7}{8} - 1\dfrac{3}{8} =$ _____

13. $\dfrac{5}{6} - \dfrac{2}{6} =$ _____

14. $\dfrac{8}{11} - \dfrac{4}{11} =$ _____

15. $10\dfrac{5}{7} - 4\dfrac{3}{7} =$ _____

Solve the following word problems.

16. Melissa brought all of her hair ribbons to Amy's house. She gave $\frac{2}{4}$ of her ribbons to Amy. How many ribbons did Melissa have left? (Hint: The total amount of ribbons to begin with is $\frac{4}{4}$.)

17. John Henry was saving up for a new baseball glove. It will take 5 weeks in all to save enough money. John Henry has been saving for 3 weeks which brings him $\frac{3}{5}$ of the way to his goal. What fraction does John Henry have left to go to reach his goal?

Change the mixed numbers into improper fractions.

18. $6\dfrac{2}{3} =$ _____

19. $2\dfrac{5}{7} =$ _____

20. $4\dfrac{3}{4} =$ _____

21. $5\dfrac{1}{2} =$ _____

Change the improper fractions to mixed numbers.

22. $\dfrac{11}{3} =$ _____

23. $\dfrac{17}{5} =$ _____

24. $\dfrac{22}{4} =$ _____

25. $\dfrac{42}{6} =$ _____

Chapter 3 Test

1. Change the improper fraction to a mixed number.

$$\frac{31}{6}$$

A. $5\frac{1}{6}$

B. $5\frac{3}{6}$

C. $5\frac{2}{6}$

D. $6\frac{1}{6}$

2. Change $4\frac{3}{7}$ to an improper fraction.

A. $\frac{30}{7}$

B. $\frac{14}{7}$

C. $\frac{31}{7}$

D. $\frac{10}{4}$

3. Add: $\frac{3}{8} + \frac{2}{8}$

A. $\frac{1}{8}$

B. $\frac{32}{8}$

C. $\frac{6}{8}$

D. $\frac{5}{8}$

4. Subtract: $\frac{9}{12} - \frac{5}{12}$

A. $\frac{4}{12}$

B. $\frac{14}{12}$

C. $4\frac{1}{12}$

D. $4\frac{4}{12}$

5. What fraction is the shaded part?

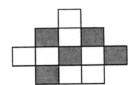

A. $\frac{4}{12}$

B. $\frac{5}{12}$

C. $\frac{5}{16}$

D. $\frac{4}{16}$

6. Add: $\frac{2}{11} + \frac{7}{11}$

A. $\frac{27}{11}$

B. $\frac{8}{11}$

C. $\frac{9}{11}$

D. $\frac{10}{11}$

7. Change the mixed number $6\frac{3}{5}$ to an improper fraction.

   A. $\dfrac{33}{5}$

   B. $\dfrac{9}{5}$

   C. $\dfrac{3}{5}$

   D. $\dfrac{30}{5}$

8. Terry has eaten $\frac{1}{6}$ of a cherry pie. Her dad ate $\frac{2}{6}$ of the cherry pie. How much of the cherry pie has been eaten?

   A. $\dfrac{3}{4}$

   B. $\dfrac{3}{5}$

   C. $\dfrac{3}{12}$

   D. $\dfrac{3}{6}$

9. Change the improper fraction $\dfrac{17}{4}$ to a mixed number.

   A. $4\frac{1}{4}$

   B. $4\frac{2}{4}$

   C. $3\frac{2}{4}$

   D. $4\frac{3}{4}$

10. Wayne's big bag of marbles weighs $\frac{5}{8}$ of a pound and his smaller bag of marbles weighs $\frac{2}{8}$ of a pound. What is the total weight of the two bags of marbles?

    A. $\dfrac{3}{8}$ lb.

    B. $\dfrac{6}{8}$ lb.

    C. $\dfrac{7}{8}$ lb.

    D. $\dfrac{5}{8}$ lb.

11. Write a fraction for the part of the figure that is shaded.

    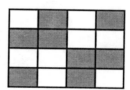

    A. $\dfrac{7}{16}$

    B. $\dfrac{6}{16}$

    C. $\dfrac{8}{12}$

    D. $\dfrac{8}{16}$

12. Add: $\dfrac{8}{11} + \dfrac{2}{11}$

    A. $\dfrac{6}{11}$

    B. $\dfrac{82}{11}$

    C. $\dfrac{10}{11}$

    D. $\dfrac{28}{11}$

13. Subtract: $\dfrac{5}{12} - \dfrac{4}{12}$

   A. $\dfrac{1}{12}$

   B. $\dfrac{9}{12}$

   C. $\dfrac{14}{12}$

   D. $\dfrac{15}{12}$

14. Change $\dfrac{51}{7}$ into a mixed number.

   A. $9\frac{2}{7}$

   B. $7\frac{2}{7}$

   C. $7\frac{5}{7}$

   D. $9\frac{5}{7}$

15. Change the improper fraction to a mixed number.

$$\dfrac{25}{8}$$

   A. $3\frac{2}{8}$

   B. $4\frac{1}{8}$

   C. $3\frac{1}{8}$

   D. $4\frac{2}{8}$

16. Convert $5\frac{2}{9}$ to an improper fraction.

   A. $\dfrac{47}{9}$

   B. $\dfrac{43}{9}$

   C. $\dfrac{37}{9}$

   D. $\dfrac{45}{9}$

17. Add: $3\frac{1}{4} + 2\frac{2}{4}$

   A. $5\frac{3}{8}$

   B. $5\frac{4}{11}$

   C. $6\frac{3}{4}$

   D. $5\frac{3}{4}$

18. Add: $6\frac{4}{10} + 2\frac{3}{10}$

   A. $4\frac{1}{10}$

   B. $8\frac{7}{10}$

   C. $12\frac{7}{10}$

   D. $8\frac{7}{20}$

19. Subtract: $8\frac{4}{6} - 2\frac{1}{6}$

   A. $6\frac{5}{12}$

   B. $10\frac{5}{6}$

   C. $10\frac{3}{6}$

   D. $6\frac{3}{6}$

20. Subtract: $5\frac{4}{11} - 3\frac{3}{11}$

   A. $8\frac{7}{11}$

   B. $8\frac{1}{11}$

   C. $2\frac{1}{11}$

   D. $2\frac{7}{11}$

Chapter 4
Decimals

This chapter covers the following Georgia Performance Standards:

M4N	Numbers and Operations	M4N2.c
M4N	Numbers and Operations	M4N5.a, b, c, d,e

4.1 Reviewing Decimal Fractions

A decimal point is the point after a whole number and before the fraction of a whole number. A decimal number has two parts, a whole number part followed by a ("."), then after the "." is the fraction part. A decimal is part of the base-ten system.

Example 1:   The decimal fraction 1.25 is equal to 1 plus $\frac{25}{100}$. The 1 is to the left of the decimal point (".") and is a whole number. The 0.25 is to the right of the decimal point and is a fraction of the whole number, in this case 0.25.

Place Value		
Ones	Tenths	Hundredths
1 .	2	5

Example 2: The decimal fraction in the chart below is $4,615.37$. The "37" to the right of the decimal is equal to $\frac{37}{100}$. The 3 is equal to "3 tenths" of one whole number and the 7 is equal to "7 hundredths" of a whole number. Together they read from left to right as thirty seven hundredths.

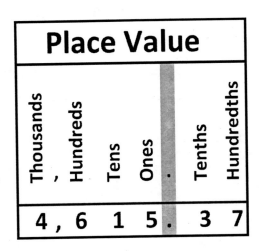

The first number to the right of the decimal is the **tenths place**.

The second number after the decimal point is the **hundredths place**.

In the example above, $4,615.37$ is equal to $4,615 + \frac{37}{100}$.

The number above in **standard** form is: $4,615.37$.

The number above in **expanded** form is: $4,000 + 600 + 10 + 5 + 0.3 + 0.07 = 4,615.37$.

The number above in **word** form is: four thousand, six hundred fifteen and thirty-even hundredths.

Remember: When writing a number in word form, use the word "and" where the decimal point goes.

An easy way to remember decimal fractions to the hundredths is to look at the way money is written.

Example 3: $\$6.28$ is a decimal fraction that equals six whole dollars and 28 cents. There are 100 cents in one dollar. In our example of $\$6.28$, we know that there are 28 cents and look how it is written–the 28 cents are written to the right of the decimal, just like a decimal fraction! 28 cents is equal to $\frac{28}{100}$ of one whole dollar. So the numbers to the right of the decimal point are less than one whole of something, in this case, a dollar.

What is the place value of the digit 8 in the following decimal numbers?

1. 0.08

2. 0.82

3. 25.8

4. 6.81

5. 0.28

6. 445.84

7. 7.38

8. $9.80

9. $14.18

Review the numbers below and write the expanded form.

10. 0.25

11. 10.68

12. 4.1

13. 0.14

14. 0.73

15. 2.16

Review the numbers below and write the word form.

16. 0.14

17. 8.71

18. 6.64

19. $4.99

20. 0.38

4.2   Ordering Decimals

Example 4:   Order the following decimals from greatest to least.

0.2, 0.02, 0.25, 0.61

Step 1:   Arrange numbers with decimal points directly under each other.

0.2
0.02
0.25
0.61

Step 2:   Fill in with zeros to the right of the number so they all have the same amount of places after the decimal point. Remember to **read the numbers as if the decimal points were not there.**

0.20
0.02 ← Least
0.25
0.61 ← Greatest

Answer: 0.02, 0.2, 0.25, 0.61

Order each set of decimals below from least to greatest.

1.   0.32, 0.03, 0.33, 0.02
2.   0.4, 0.14, 0.48, 0.18
3.   0.1, 0.01, 0.11, 0.09
4.   0.3, 0.59, 0.64, 0.12
5.   0.06, 0.88, 0.72, 0.27
6.   0.21, 0.55, 0.51, 0.15
7.   0.04, 0.09, 0.4, 0.9

8.   0.15, 0.11, 0.19, 0.17
9.   0.09, 0.99, 0.19, 0.91
10.   0.47, 0.44, 0.77, 0.74
11.   0.62, 0.06, 0.02, 0.03
12.   0.41, 0.45, 0.49, 0.43
13.   0.08, 0.07, 0.10, 0.09
14.   0.51, 0.05, 0.50, 0.15

Order each set of decimals below from greatest to least.

15.   0.13, 0.12, 0.72, 0.27
16.   0.42, 0.24, 0.99, 0.88
17.   0.01, 0.25, 0.36, 0.13

18.   0.68, 0.52, 0.86, 0.25
19.   0.28, 0.82, 0.18, 0.81
20.   0.05, 0.5, 0.45, 0.54

4.3   Adding Decimals

Example 5:   Add: $7.8 + 1.16$

Step 1:   When adding decimals, first arrange the numbers in columns with the decimal points under each other. If necessary, fill in with zeros to the right of the number, so all the numbers have the same amount of places after the decimal point.

Step 2:   Start at the right and add each column. Remember to carry when necessary. Bring the decimal point straight down in the answer. Let's add $7.8 + 1.16$.

Ones	.	Tenths	Hundreths
7	.	8	0
+ 1	.	1	6
8	.	9	6

Answer: 8.96

Add the following. Be sure to write the decimal in your answer.

1. $8.24 + 6.3$

2. $5.4 + 0.81$

3. $4.27 + 0.22$

4. $2.99 + 4.99$

5. $1.17 + 7.17$

6. $6.32 + 0.24$

7. $0.73 + 0.52$

8. $4.58 + 0.24$

9. $2.32 + 0.64$

10. $8.68 + 0.08$

11. $0.6 + 0.06$

12. $7.45 + 0.2$

4.4   Subtracting Decimals

Example 6:   Subtract: $9.45 - 6.2$

Step 1:   When you subtract decimals, first arrange the numbers in columns with the decimal points under each other. If necessary, fill in with zeros to the right of the number, so all the numbers have the same amount of places after the decimal point.

Step 2:   Subtract as you would any subtraction problem. Remember to borrow when needed. Bring the decimal point straight down in the answer. Let's subtract $9.45 - 6.2$.

Ones	.	Tenths	Hundreths
9	.	4	5
− 6	.	2	0
3	.	2	5

Answer: 3.25

Subtract the following. Be sure to write the decimal in your answer.

1. $12.1 - 3.04$

2. $9.65 - 4.21$

3. $0.22 - 0.21$

4. $0.89 - 0.05$

5. $7.72 - 2.27$

6. $11.54 - 9.65$

7. $34.05 - 2.5$

8. $4.6 - 3.02$

9. $54.54 - 45.45$

10. $10.1 - 0.11$

11. $86.32 - 27.42$

12. $12.04 - 1.09$

4.5   Multiplying Decimals by Whole Numbers

Example 7:   Multiply 21.4×5

Step 1:   Since the problem above was written horizontally, we should rewrite the problem vertically.

$$\begin{array}{r} 21.4 \\ \times \quad 3 \\ \hline \end{array}$$

Step 2:   Multiply as if all numbers are whole and carry when possible. Disregard the decimal point, for now.

$$\begin{array}{r} 21.4 \\ \times \quad 3 \\ \hline 642 \end{array}$$

Step 3:   For each factor, count the numbers that appear after the decimal point.

$$\begin{array}{r} 21.4 \\ 3 \\ \hline \end{array} \quad \begin{array}{l} \text{1 number after the decimal} \\ +0 \text{ numbers after the decimal} \\ \hline \text{1 number after the decimal in total} \end{array}$$

Step 4:   If there is one number after the decimal point in the problem, there should be one number after the decimal point in the answer.
Product should have one $(1 + 0)$ number after the decimal: 64.2

Answer: 64.2

Rewrite the problem vertically and multiply.

1. 2.3×7

2. 6.82×4

3. 3.21×18

4. 45.37×8

5. 61.09×19

6. 7.3×12

7. 48.55×33

8. 18.18×18

9. 27.04×62

10. 3.7×122

11. 9.97×3

12. 64.29×87

13. 55.48×24

14. 6.89×61

15. 74.33×56

16. 84.2×45

17. 18.1×44

18. 7.4×39

4.6 Division of Decimals by Whole Numbers

Example 8: $48.36 \div 12$

Step 1:   Copy the problem as you would for whole numbers. Copy the decimal point directly above in the place for the answer (the quotient).

$$12{\overline{)48.36}}$$

Step 2:   Divide the same way as you would with whole numbers.

```
         4.03
  12)48.36
     -48 ↓|
       03 |
        0 ↓
        36
        36
         0
```

Divide. Remember to copy the decimal point directly above the place for the answer.

1. $12.36 \div 2$

2. $55.5 \div 5$

3. $27.09 \div 9$

4. $63.3 \div 3$

5. $100.1 \div 10$

6. $81.9 \div 9$

7. $44.12 \div 4$

8. $500.25 \div 5$

9. $640.8 \div 8$

10. $25.46 \div 2$

11. $639.39 \div 3$

12. $28.7 \div 7$

13. $88.11 \div 11$

14. $42.28 \div 14$

15. $999.99 \div 9$

16. $121.11 \div 11$

4.7   Estimating Decimals Using Rounding

Example 9:   Estimate 8.5 to the nearest whole number.

Step 1:   Find the whole number and the decimal fraction of 8.5. The whole number is the "8", in the ones place, and the decimal fraction is the "5" in the tenths place.

Step 2:   Look at the number of the tenths place. If the number in the tenths place is 5 or greater, then the number in the ones column will round up to the next whole number. If the number in the tenths place is less than 5, the number will round down and the whole number in the ones column, will remain the same.

Step 3:   Because the number in the tenths place is 5, we round the whole number from 8 to 9.

Answer: 9

Example 10:   Estimate 12.87 to the nearest whole number.

Step 1:   The whole number is 12. The decimal fraction is 87 hundredths.

Step 2:   Look at the number directly right of the decimal. Since 8 is greater than 5, round up.

Step 3:   The whole number increases by one from 12 to 13.

Answer: 13

Round the decimal fractions to the nearest whole number.

1.	11.4	11.	984.06	21.	7,989.3
2.	278.9	12.	4,656.7	22.	8,251.05
3.	56.31	13.	489.89	23.	77.77
4.	12.55	14.	753.2	24.	22.22
5.	1,458.7	15.	15.83	25.	64.6
6.	23.62	16.	87.2	26.	82.03
7.	9.91	17.	32.54	27.	94.1
8.	147.47	18.	46.72	28.	335.51
9.	83.2	19.	121.21	29.	642.6
10.	347.6	20.	17.4	30.	145.05

4.8   Decimal Word Problems

Example 11:   Gilbert buys 5 boxes of cookies for $3.75 each. How much in total did her spend?

Step 1:   Determine the operation. Each box costs $3.75, and we want to find the total. We must multiply.

Step 2:   Multiply: $3.75 × 5.

$$
\begin{array}{r}
{\scriptstyle 3\ 2} \\
\$3.75 \\
\times\ \ \ \ 5 \\
\hline
\$18.75
\end{array}
$$

Answer: $18.75

Answer the following multiplication and division problems.

1. Henry wants to buy two movies with his birthday money. Each movie costs $19.99. How much money does Henry need to buy the movies?

2. Woodstock Middle School spends $1,443.20 on 55 math books. How much does each book cost?

3. Ms. Laudry needs 8 bags of flour for her bakery. Each bag of flour weighs 3.25 pounds. How many pounds of flour does Ms. Laudry need?

4. Andrea has 17.28 pounds of candy. She is going to divide this candy among herself and three friends. How many pounds of candy will each person get?

5. Anthony spends $243.18 on three new bikes for his three daughters. How much did each bike cost?

6. A family of five is going to the movie theater. If each ticket costs $8.75. How much will it cost this family to see a movie if all of the tickets are the same price?

Chapter 4 Review

Write the place value of the digit 7 in the following decimal fractions.

1. 298.71

3. 17.35

5. 99.7

2. 15.27

4. 86.67

6. 2.07

Write the numbers below in expanded form.

7. 12.7

8. 6.54

9. 2.02

Order each set of decimals below from least to greatest.

10. 0.27, 5.27, 0.07, 0.77

11. 4.84, 8.45, 7.08, 4.48

12. 0.33, 3.33, 0.03, 0.3

Add the following. Be sure to write the decimal in your answer.

13. $6.72 + 0.23$

15. $8.9 + 0.04$

17. $0.34 + 1.12$

14. $4.55 + 2.47$

16. $6.89 + 0.63$

18. $5.5 + 0.77$

Subtract the following. Be sure to write the decimal in your answer.

19. $9.21 - 0.25$

21. $45.27 - 0.19$

23. $55.55 - 22.22$

20. $14.68 - 0.03$

22. $28.3 - 0.14$

24. $7.43 - 1.28$

Multiply.

25. 4.2×3

27. 3.26×4

29. 2.71×4

26. 22.1×12

28. 8.33×3

30. 44.81×3

Divide.

31. $52.26 \div 13$

33. $70.5 \div 5$

35. $64.32 \div 8$

32. $33.33 \div 11$

34. $18.63 \div 9$

36. $72.96 \div 12$

Round the following decimal fractions to the nearest whole number.

37. 62.7

39. 1,665.2

41. 8,991.45

38. 83.04

40. 62.77

42. 6.06

Chapter 4 Test

1. What is the place value of the digit 8 in the number 961.85?

   A. Eight hundred
   B. Eight hundredths
   C. Eight tens
   D. Eight tenths

2. Put the following decimal fractions in order from least to greatest.
   4.52, 0.45, 5.24, 2.45

   A. 0.45, 2.45, 4.52, 5.24
   B. 5.24, 4.52, 2.45, 0.45
   C. 2.45, 0.45, 4.52, 5.24
   D. 4.52, 5.24, 0.45, 2.45

3. Add: $62.45 + 22.31$

   A. 62.76
   B. 84.76
   C. 40.14
   D. 62.67

4. Subtract: $87.23 - 55.62$

   A. 30.61
   B. 142.85
   C. 31.61
   D. 141.84

5. Multiply: 66.36×4

   A. 265.44
   B. 241.44
   C. 254.65
   D. 256.44

6. Divide: $66.33 \div 3$

   A. 20.11
   B. 21.11
   C. 22.00
   D. 22.11

7. Round 844.36 to the nearest whole number.

   A. 844.3
   B. 845
   C. 844
   D. 844.4

8. What place value is the 2 in 645.12?

   A. tenths
   B. hundredths
   C. tens
   D. hundreds

9. Add: $22.22 + 87.99$

   A. 111.21
   B. 109.22
   C. 110.22
   D. 110.21

10. Multiply: 11.11×12

    A. 133.32
    B. 111.11
    C. 122.12
    D. 132.12

11. Subtract: $742.83 - 12.47$

    A. 733.36
    B. 730.36
    C. 730.41
    D. 733.41

12. Divide: $9.4 \div 2$

    A. 7.4
    B. 4.4
    C. 4.7
    D. 7.2

13. Put in order from greatest to least.
    0.12, 1.12, 2.11, 0.22

    A. 2.11, 1.12, 0.22, 0.12
    B. 1.12, 0.22, 2.11, 0.12
    C. 0.12, 0.22, 1.12, 2.11
    D. 0.12, 1.12, 2.11, 0.22

14. Add: $88.91 + 12.37$

    A. 100.18
    B. 101.21
    C. 101.18
    D. 101.28

15. Round 6.75 to the nearest whole number.

    A. 6
    B. 7
    C. 6.7
    D. 6.8

16. Divide: $12.48 \div 4$

    A. 4.12
    B. 3.14
    C. 3.12
    D. 4.13

17. Write 5.24 in expanded form.

    A. five and twenty four
    B. $5 + 0.2 + 0.04$
    C. five and twenty four hundredths
    D. $50 + 0.2 + 0.04$

18. Write 80.76 in word form.

    A. eighty and seventy-six hundredths
    B. eighty and seven tenths and 6 hundredths
    C. eight thousand seventy six
    D. eighty and $\frac{76}{100}$

19. Multiply: 21.56×24

    A. 514.74
    B. 517.44
    C. 456.56
    D. 517.64

20. Subtract: $4,627.67 - 1.09$

    A. 4,626
    B. 4,626.58
    C. 4,627.58
    D. 4,626.76

Chapter 5
Arithmetic Operations

This chapter covers the following Georgia Performance Standards:

M4N	Numbers and Operations	M4N7.a, b, c, d

5.1 Using Addition, Subtraction, Multiplication, and Division

In order to solve math problems in every day life, it is important to know whether to add, subtract, multiply, or divide the numbers involved. Addition, subtraction, multiplication, and division are called **arithmetic operations**. In order to choose the correct operation, you need to read the problem very carefully and figure out the goal of the problem.

Example 1: Henry has two female dogs named Honey Bun and Sugar-Pie. Both dogs gave birth to puppies in the last week. Honey Bun had 6 puppies and Sugar-Pie had 8 puppies. What is the total number of puppies Honey Bun and Sugar-Pie have?

Step 1: Figure out the goal of the problem. In this case, we want to know the total number of puppies born to Honey Bun and Sugar-Pie. The way to find the total of something is to use addition.

Step 2: Add the total number of puppies: $6 + 8 = 14$.

Answer: 14 puppies

Below is a chart of some examples of key words and phrases used in word problems for each operation.

Operation	Key Words and Phrases
Addition	Total of, Sum of, In all, Plus, Add
Subtraction	Remaining, Are left, Minus, Take away
Multiplication	Times, How much of many items, Multiply
Division	Equal parts, In each set, Divided equally

Example 2: Mrs. Lundgrin's 4th grade class is going on a field trip to the circus. The total cost for the 32 students plus Mrs. Lundgrin came to $330.00. What is the cost per person, if the total cost is divided equally among the students and Mrs. Lundgrin?

Step 1: Figure out the goal of the problem. In this case we want to know how much the cost of the ticket is per person. Using the chart on the previous page, let's find a key phrase used in the problem to help us figure out which operation to use. We see "divided equally" used, so division is the operation we will use.

Step 2: Set up the math problem. $330.00 (the total cost) ÷ 33 (the total number of people going to the circus) = ?

$$330.00 ÷ 33 = ? \qquad\qquad 330.00 ÷ 33 = 10.00$$

Answer: $10.00 per person.

Solve the problems below.

1. Karitha wanted to buy a CD of her favorite music group. The CD, including tax, costs $16.50. Karitha has already saved $12.75 to use for the CD purchase. What is the remaining amount of money Karitha will need to save up before she can purchase the CD?

2. Henry owes his stepfather $20.00. If Henry pays his stepfather back in payments over 4 weeks, how much will each weekly payment need to be if Henry makes 4 equal payments?

3. Lila has 23 white buttons, 14 red buttons, 6 green buttons, 8 blue buttons and 12 brown buttons in her button collection. How many buttons in all does Lila have in her button collection?

4. Ching Ngo was making egg rolls to sell at the school bake sale. Ching makes the egg rolls in batches of 24 and she has made 4 batches so far. How many egg rolls has Ching made so far?

5. Jeremy found after eating 2 pieces of fruit every day for a week, he felt better and had more energy. How many pieces of fruit will Jeremy eat in 57 days if he keeps eating 2 pieces of fruit every day?

6. Stanley needs to earn $12.00 to buy his mother a gift for her birthday. Stanley already has $9.80. If Stanley takes away the $9.80 he already has from the $12.00 he needs, how much more money does he need to earn to buy his mother's present?

5.2 Order of Operations

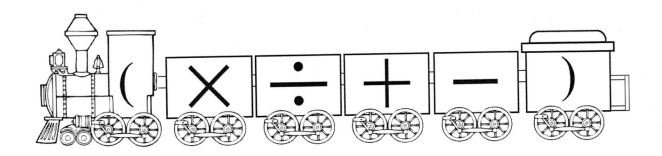

There are some math problems that use more than one arithmetic operation. Some of these longer math problems use parentheses that look like this (). Inside the parentheses are smaller math problems to use within the larger math problem. All of the operations must be done in a certain order for the answer to be correct. Does it sound complicated? Just keep my Aunt Sally happy and it will all work out! What do I mean by that? Try to remember the phrase **Please My Dear Aunt Sally** and you will know the order that arithmetic operations should be done. This is how you simplify the order of operations. See below and you'll understand.

Please "P" stands for parentheses. To simplify the problem, first you must get rid of the parentheses, by doing the smaller math problem within the parentheses.
Examples: $2 \times (4 - 1) = 2 \times 3 = 6$ $8 \div (12 - 8) = 8 \div 4 = 2$

My Dear "M" stands for multiply. "D" stands for divide. Start on the left of the problem and do all multiplications and divisions in the order in which they appear.

Aunt Sally "A" stands for add. "S" stands for subtract. Start on the left and do all additions and subtractions in the order they appear.

Example 3: $5 \times (9 - 3) + 12 =$

Please Eliminate **parentheses**. $9 - 3 = 6$ so now we have $5 \times 6 + 12$

My Dear **Multiply** and **divide** next in order from left to right. $5 \times 6 = 30$, then we have $30 + 12$

Aunt Sally Last, we **add** and **subtract** in order from left to right. $30 + 12 = 42$

Simplify and solve the problems below.

1. $8 + (2 + 3) - 12$

2. $2 \times (7 - 2) + 12$

3. $9 - (6 - 2)$

4. $50 \div (15 - 5)$

5. $62 + (18 \div 3) + 4$

6. $17 - 5 + (8 \times 2)$

7. $83 + (6 - 4) + 5$

8. $3 \times (8 - 4) + 11$

9. $19 - 2 + (9 \div 3)$

10. $50 \div 2 + (7 - 2)$

11. $6 + (37 \times 2) - 14$

12. $4 \times (8 - 3) + 20$

13. $36 \div 6 + (19 - 16)$

14. $2 + 3 - (87 - 84)$

15. $45 \div 9 + (16 \div 4)$

16. $100 - 95 + (49 \div 7)$

17. $(650 \div 50) + 19$

18. $597 - 582 + (6 \times 2)$

19. $1 + 4 + 5 + (2 \div 2)$

20. $25 + (25 \div 5) + 5$

5.3   Properties of Mathematics

There are three properties of mathematics that can be used to solve addition and multiplication problems. Properties are rules for arithmetic operations. The chart below names each property and explains how they work. These properties only work for addition and multiplication; do not use them for subtraction or division. You will get the wrong answer!

Property	Rule of the Property	Example
Commutative	If you change the order of the factors, the product is the same.	$4 \times 3 = 3 \times 4$ $12 = 12$ $6 + 8 = 8 + 6$ $14 = 14$
Associative	When adding or multiplying 3 numbers, the solution will be the same no matter which order you group the numbers.	$2 \times (3 \times 4) = (2 \times 3) \times 4$ $24 = 24$ $3 + (5 + 4) = (3 + 5) + 4$ $12 = 12$
Distributive	When multiplying one number by the sum of two other numbers, you can multiply that number outside the parentheses by each number in the parentheses.	$6 \times (2 + 1) = (6 \times 2) + (6 \times 1)$ $6 \times 3 = 12 + 6$ $18 = 18$

Identify the properties of the equations below by using <u>C</u> for Commutative, <u>A</u> for Associative, or <u>D</u> for Distributive. You do not need to solve the problems.

1. $4 \times (8 + 3) = (4 \times 8) + (4 \times 3)$

2. $6 \times 8 = 8 \times 6$

3. $5 \times (7 \times 2) = (5 \times 7) \times 2$

4. $253 \times 678 = 678 \times 253$

5. $112 \times (85 + 27) = (112 \times 85) + (112 \times 27)$

6. $745 + 999 = 999 + 745$

7. $8 + (16 + 15) = (8 + 16) + 15$

8. $912 + 455 = 455 + 912$

9. $87 \times (2 \times 6) = (87 \times 2) \times 6$

10. $83 + 17 = 17 + 83$

5.4   Using Mental Math and Estimation to Solve

Sometimes when you do math in your head, it is helpful to estimate the problem, as long as it's not critical to have the exact answer. For instance, you may only need to estimate the amount of corn seed needed to plant 175 acres of corn, but knowing the exact amount for a scientific experiment may be crucial!

Example 4:   Sandra needs 6 yards of fabric for each choir robe for the Elementary School of Music. There are 24 students in the school choir. Sandra is at the fabric store and needs to figure out in her head how much fabric to buy for all 24 students. To make it easier, she can use the distributive property to figure the total number of yards needed.

Step 1:   Sandra needs 6 yards × 24 robes. If Sandra splits up the number 24 into 20 + 4, she will make it easier to multiply in her head.

$$\begin{aligned} 6 \times (20 + 4) &= (6 \times 20) + (6 \times 4) \\ &= 120 + 24 \\ &= 144 \end{aligned}$$

Answer: Sandra needs 144 yards of fabric.

Example 5:   Chad lives on a tight budget and is trying to figure out which is a better deal on dog food for his two hound dogs. One package is 3 pounds for a total of $2.97 and the other package is 10 pounds for $8.92. What is the easiest way for Chad to figure the best buy in his head?

Step 1:   Chad can estimate the cost per pound by rounding the price to the nearest dollar and then dividing the number of dollars by the number of pounds. The package with the lowest price per pound is the best deal.

$2.97 rounds up to $3.00 and $8.92 rounds up to $9.00.

$3.00 ÷ 3 pounds = $1.00 and $9.00 ÷ 10 pounds = $0.90

Answer: The 10 pound bag is cheapest at about $0.90 per pound.

Estimate the answers for the following problems in your head.

1. Which is a better deal? Buying 6 rolls of paper towels for $5.79 or 12 rolls of paper towels for $10.99?

2. Jeff is buying 12 packages of hotdogs at $3.98 per package for a family barbecue. About how much are the 12 packages going to cost Jeff?

3. Amy bought a three pound package of hamburger at the Greenbuy Grocer for $11.88. When she went home, she noticed a sign on the window at the Cheapway Grocer that said 5 pounds of hamburger for $18.88. Did she buy the hamburger at the best price? (Yes or no.)

Chapter 5 Review

Solve the following problems.

1. Carlos has a yo-yo collection that contains 12 red yo-yos, 4 green ones, 7 orange ones and 9 yellow ones. How many yo-yos in all does Carlos have?

2. Evan wants to evenly divide 42 pieces of candy among 20 students and 1 teacher. Including Evan, how many pieces of candy will each person receive?

3. There are 18 dog bones in a bag. Jake has given his dog 10 bones this month from the bag. How many dog bones are left in the bag?

Simplify and solve the problems below.

4. $9 + (6 - 4) + 2$

5. $(45 \div 9) - 2 + 8$

6. $4 \times (8 + 3)$

7. $2 + 9 + 3 - (2 \times 4)$

8. $16 + (9 - 5)$

9. $(27 \div 9) + 14$

10. $2 + 8 + 3 + (18 - 14)$

11. $18 \div 3 + (4 \times 2)$

12. $9 - 5 + (7 - 4)$

13. $3 \times (54 - 42)$

14. $3 + (8 + 11) + 6$

15. $99 - 88 + (6 + 2)$

16. $7 + 4 - 2 + 5 + (7 - 3)$

Identify the properties of the equations below by using <u>C</u> for Commutative, <u>A</u> for Associative, and <u>D</u> for Distributive. You do not need to solve the problems.

17. $12 \times 54 = 54 \times 12$

18. $5 + (3 + 2) = (5 + 3) + 2$

19. $8 \times (5 + 7) = (8 \times 5) + (8 \times 7)$

Using mental math, estimate and solve the following.

20. Mr. Stevens bought 6 new basketballs at the Keep-On-Movin' sporting goods store for a total of $54.00. Mrs. Stevens noticed an ad for basketballs at the local mart store for $7.99 each. Did Mr. Stevens get the best deal on the basketballs?

21. Carrie needs two new pair of shoes. The department store has the shoes Carrie wants, a pair of black dress shoes for $14.99 and a pair of sneakers for $15.99. The Allgood Shoe Store prices their shoes much higher, but they are having a "buy one, get one free" sale. If the shoes at the Allgood Shoe Store sell for $24.99 each, would it be a better deal for Carrie to buy her shoes at the department store or at the Allgood Shoe Store, while the sale is still on?

22. Bobbie and Jamie have decided to trade trumpets. Bobbie has played his trumpet for 8 hours every week since school started, 22 weeks ago. Jamie has played his trumpet 12 hours every week since his uncle bought it for him 15 weeks ago. Which boy has played more hours on his trumpet?

Chapter 5 Test

Solve the following problems.

1. In the boxed set of dishes that Lynda wants to buy, there are 4 each of dinner plates, bowls, salad plates, cups, and saucers. How many dishes are in the box?

   A. 25
   B. 15
   C. 20
   D. 9

2. Thaddeus had $27.65 in his piggy bank last week. Since then, Thaddeus has purchased a book for $11.95. How much money does Thaddeus have remaining?

   A. $15.70
   B. $39.60
   C. $15.30
   D. $16.70

3. Erin was told by her teacher, Mrs. Lightheart, to pass out all the crayons to the 20 students in class. At last count, there were 140 crayons in the box. How many crayons will each student receive?

   A. 8
   B. 7
   C. 6
   D. 9

Simplify and solve the problems below.

4. $4 + 4 + 4 + (4 \times 4)$

   A. 28
   B. 20
   C. 24
   D. 16

5. $7 - (19 - 15)$

   A. 15
   B. 3
   C. 27
   D. 9

6. $(35 \div 5) + 13$

   A. 25
   B. 15
   C. 20
   D. 9

7. $2 \times (18 \div 2) + 3$

   A. 14
   B. 18
   C. 20
   D. 21

8. $14 - 3 + 2 + (8 - 3)$

A. 18
B. 17
C. 28
D. 13

9. $6 + 2 - 7 + 12 + (8 \div 4)$

A. 16
B. 15
C. 17
D. 18

10. $27 \div (7 - 4)$

A. 16
B. 18
C. 11
D. 9

11. $2 + 12 + 14 - 11 - 3$

A. 16
B. 15
C. 14
D. 13

12. $(8 \times 3) - 20 - 3$

A. 1
B. 2
C. 11
D. 12

13. $1 + 1 + 1 - 1 + (1 \times 1) - 1$

A. 4
B. 2
C. 3
D. 1

Identify the properties for the equations below. Use <u>C</u> for Commutative, <u>A</u> for Associative, and <u>D</u> for Distributive. You do not need to solve the problems.

14. $652 \times 555 = 555 \times 652$

A. C
B. A
C. D
D. None of the above

15. $4 + (12 + 11) = (4 + 12) + 11$

A. A
B. C
C. D
D. None of the above

16. $8 \times (77 + 99) = (8 \times 77) + (8 \times 99)$

A. A
B. C
C. D
D. None of the above

17. $6 \times 15 = 15 \times 6$

A. C
B. A
C. D
D. None of the above

Estimate the answers for the following problems in your head.

18. Mr. Wallace wants to grow a lot of green beans (his grandson's favorite) in his garden. If for every fifteen-foot row, Mr. Wallace needs 30 seeds, how many seeds does he need for 4, thirty-foot rows?

    A. 240 seeds
    B. 360 seeds
    C. 60 seeds
    D. 120 seeds

19. Mike had a pet rat, Jeeves, that he taught to run a maze. Jeeves received a treat for successfully completing the maze. Jeeves could run the maze from start to finish in 32 seconds. Mike bought a new rat, Lionel, and it took him 64 seconds to run the maze his first time. Which of the following statements is true?

    A. Lionel runs the maze twice as fast as Jeeves.
    B. Jeeves runs the maze twice as fast as Lionel.
    C. Jeeves and Lionel cannot be compared.
    D. Lionel and Jeeves ran the maze at the same rate.

20. Donna wants to make sure she has enough sugar in the house to make 5 batches of cookies for the school bake sale. It takes 2 cups of sugar per batch. Each box of sugar contains 4 cups and Donna has 3 full boxes of sugar. Does she have enough sugar to make 5 batches of cookies?

    A. Donna needs 10 cups of sugar and she has 12 cups. Yes, she has enough sugar.
    B. Donna needs 12 cups of sugar and she has 10 cups. She needs one more box of sugar.
    C. Donna needs 10 cups of sugar and she has 14 cups. Yes, she has enough sugar.
    D. Donna needs 12 cups of sugar and she has 6 cups. She needs two more boxes of sugar.

We're done with the test!

Chapter 6
Mathematical Relationships in Algebra

This chapter covers the following Georgia Performance Standards:

| M4A | Algebra | M4A1.a, b, c |

6.1   Patterns

A **pattern** is a set of like items such as numbers or shapes that are repeated in a certain order. Patterns are used in almost every kind of work, whether it's graphic design patterns used on fabric, musical patterns used in song writing, plant designs used in landscaping or farming, architectural design for buildings and communities, or the patterns and rules of bookkeeping, and so on.

Patterns can be represented by a series of shapes or items. Repeating the series in the same order defines the pattern.

Example 1:   Define the pattern below.

Step 1:   Identify the pattern. We begin with a vertical striped box, followed by a polka-dotted box, followed by a swirled box, followed by a horizontal striped box. Then the pattern begins to repeat with a vertical striped box, followed by a polka-dotted box.

Step 2:   Which box will come next? If we study the pattern carefully, we can see that the polka-dotted box is followed by a swirled box. The swirled box comes next. The swirled box will be followed by the horizontal striped box and so on.

Example 2:   Which symbol will come next in the pattern?

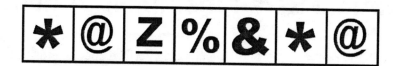

Step 1:   Identify the pattern. We can see that the first five symbols make up the pattern, because the [*] and the [@] are repeating the first two blocks of the pattern.

Step 2:   What comes next in the pattern? As we can see, the [Z] comes next, followed by the [%].

Fill in the blanks in the following patterns.

1.

2.

3.

4.

5.

6.2   Patterns Using the Four Mathematical Operations

Number patterns using addition and subtraction are similar to the patterns from the previous page. Instead of symbols, we use numbers to make the patterns. For instance, counting by two's is a number pattern using addition. Or starting from 100 and counting backwards by fours is a number pattern using subtraction.

Example 3:   What number would continue the pattern $1, 3, 5, 7, \_\_\_$?

Step 1:   In this pattern, the numbers are increasing by 2. Therefore, the next number in the pattern would be **9**.

Answer: 9

Example 4:   What number is missing in this pattern? $30, 25, 20, \_\_\_, 10, 5$.

Step 1:   In this pattern, the numbers are decreasing by 5. Therefore, the number missing in the pattern is **15**.

Answer: 15

Find the missing number in the following patterns using addition and subtraction.

1. $2, 4, 6, \_\_\_, 10, 12$

2. $100, 90, 80, 70, \_\_\_$

3. $60, 63, 66, \_\_\_, 72, 75$

4. $17, 15, 13, 11, \_\_\_, 7$

5. $40, 44, \_\_\_, 52, 56, 60$

6. $18, 23, 28, \_\_\_, 38, 43$

7. $93, 87, 81, 75, \_\_\_, 63$

8. $11, 19, \_\_\_, 35, 43, 51$

9. $62, 59, 56, 53, \_\_\_, 47$

10. $55, 53, 51, \_\_\_, 47, 45$

11. $32, 42, \_\_\_, 62, 72, 82$

12. $300, 250, 200, \_\_\_, 100$

13. $55, 66, \_\_\_, 88, 99, 110$

14. $220, 210, 200, 190, \_\_\_$

15. $717, 723, 729, \_\_\_, 741$

16. $1, 40, 79, \_\_\_, 157, 196$

17. $487, 474, 461, 448, \_\_\_$

18. $22, 62, \_\_\_, 142, 182, 222$

19. $18, 16, 14, 12, \_\_\_, 8$

20. $388, 390, \_\_\_, 394, 396$

Number patterns using multiplication and division are a little more challenging. If the pattern isn't obvious, you will need to figure it out by trial and error. But there are clues you can use to determine if the pattern is using multiplication or division. If the pattern is increasing, you will use multiplication. If the pattern is decreasing, you will use division.

Example 5:   What number would continue the pattern $2, 6, 18, 54, \underline{\quad}$?

  Step 1:   In this pattern, the numbers are increasing by a factor of 3. Therefore, the next number in the pattern would be $54 \times 3 = 162$.

Answer: **162**

Example 6:   What number is missing in the pattern $1000, 100, \underline{\quad}, 1$?

  Step 1:   In this pattern, the numbers are decreasing – dividing by 10. Therefore, the missing number in the pattern is $100 \div 10 = 10$.

Answer: **10**

Find the missing number in the following patterns using multiplication and division.

1. $125, 25, \underline{\quad}, 1$

2. $2, 4, 8, \underline{\quad}, 32, 64$

3. $80, 40, 20, \underline{\quad}, 5$

4. $3, 12, \underline{\quad}, 192, 768$

5. $640, 80, \underline{\quad}$

6. $11, \underline{\quad}, 1331, 14641$

7. $1000, 500, \underline{\quad}, 125$

8. $200, 400, 800, \underline{\quad}$

9. $3, 6, 12, \underline{\quad}, 48, 96$

10. $4, 12, \underline{\quad}, 108, 324$

11. $6, 12, 24, \underline{\quad}, 96$

12. $640, 160, \underline{\quad}, 10$

13. $1, 7, \underline{\quad}, 343, 2401$

14. $240, 120, \underline{\quad}, 30, 15$

15. $3, 9, 27, \underline{\quad}, 243$

16. $1, 1, 1, \underline{\quad}, 1, 1, 1$

17. $2, 10, \underline{\quad}, 250, 1250$

18. $48, \underline{\quad}, 12, 6, 3$

19. $30, 60, \underline{\quad}, 240, 480$

20. $1600, 400, 100, \underline{\quad}$

6.3   Open Number Sentences

An **open number sentence** is a number sentence with a number missing. Your job is to find the missing number.

Example 7:   Find the missing number.

$$15 + \boxed{} = 23$$

Step 1:   The square represents the missing term. We need to find a number that added to 15 will equal 23.

Step 2:   If we take the largest number, 23, and subtract the known number, 15, we will get the unknown number. $23 - 15 = 8$.

Step 3:   Fill in your answer, 8, in the square. Double check your answer by making sure the number sentence is correct. Does $15 + 8 = 23$? Yes, it does. The correct answer is 8.

Answer: 8

Example 8:   Find the missing number.

$$100 \div \triangle = 4$$

Step 1:   The triangle represents the missing term. We need to find a number that 100 can be divided by to equal 4.

Step 2:   If we divide 100 by 4, we will get the missing term. $100 \div 4 = 25$.

Step 3:   Fill in the answer, 25, in the triangle. Double check your answer by making sure the number sentence is correct. Does $100 \div 25 = 4$? Yes, it does. The correct answer is 25.

Answer: 25

Some open number sentences may be missing more than one term. If the missing numbers are equal, the symbol representing the missing number will be identical. If they are different numbers missing from the same open number sentence, the symbols representing the missing numbers will be different, such as $\square \times \square = 49$ (answer is 7)   or $\triangle + \square = 4$ (answer is 1 and 3).

Find the missing term(s). If there are two open number sentences separated by a comma, be certain that the numbers for each symbol will fit both sentences.

1. $8 \times \square = 24$

2. $\triangle \div 10 = 10$

3. $110 - \triangle = 60$

4. $\square + 16 = 50$

5. $\triangle + \triangle = 88$

6. $254 - \square = 112$

7. $144 \div \triangle = 12$

8. $\square \times 9 = 81$

9. $\triangle \times \triangle = 36$

10. $256 \div 2 = \square$

11. $90 - \triangle = 87$

12. $\square + \square = 8, \square \times 3 = 12$

13. $5 \times \triangle = 15, \triangle + \triangle = \square$

14. $\triangle \div 12 = 60$

15. $27 + 13 = \square$

16. $\triangle + 82 = 117$

17. $6 + \square = 18, \square \times 4 = \triangle$

18. $48 \div 8 = \square$

19. $945 - \triangle = 32$

20. $70 \div \square = 10$

21. $\triangle \times 40 = 240, \triangle - 3 = 3$

22. $777 - \square = 222$

23. $\square \times \square = 121, \square \times 3 = 33$

24. $\triangle + \square = 17, \triangle - \square = 15$

Chapter 6 Review

Fill in the blanks of the following patterns.

1.

2.

3.

Find the missing number in the following patterns using addition, subtraction, multiplication and/or division.

4. $4, 8, 12, \_\_\_, 20, 24$

9. $25, 30, \_\_\_, 40, 45$

5. $82, 76, 70, \_\_\_, 58$

10. $101, 91, 81, 71, \_\_\_$

6. $64, 16, \_\_\_, 1$

11. $4, 8, 16, \_\_\_, 64, 128$

7. $3, 9, 27, \_\_\_, 243$

12. $23, 30, 37, \_\_\_, 51$

8. $81, 27, 9, \_\_\_$

13. $914, 903, 892, 881, \_\_\_$

Find the missing term.

14. $16 \div \triangle = 4$

19. $83 - \square = 27$

15. $24 \times \square = 48$

20. $\triangle \times 12 = 60$

16. $\triangle - 30 = 70$

21. $888 + \square = 999$

17. $45 + \square = 94$

22. $100 \div \triangle = 4$

18. $36 \div 9 = \triangle$

23. $15 \times \square = 60$

Chapter 6 Test

Find the missing figure in the pattern.

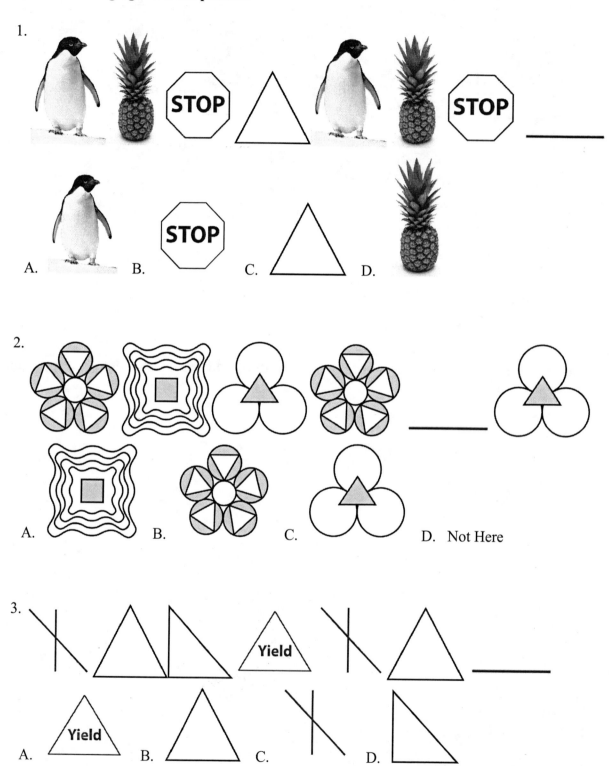

1.

A.   B.   C.   D.

2.

A.   B.   C.   D. Not Here

3.

A.   B.   C.   D.

Find the missing number in the following patterns using addition, subtraction, multiplication and division.

4. 812, 808, 804, ___, 796, 792

   A. 800
   B. 801
   C. 802
   D. 799

5. 64, 16, ___, 1

   A. 8
   B. 2
   C. 4
   D. 12

6. 15, 45, ___, 405, 1215

   A. 120
   B. 140
   C. 145
   D. 135

7. 7, 14, 21, 28, ___, 42

   A. 32
   B. 36
   C. 35
   D. 30

8. 542, 537, ___, 527, 522

   A. 530
   B. 532
   C. 534
   D. 531

9. 999, 333, 111, ___

   A. 37
   B. 3
   C. 33
   D. 13

10. 17, 20, 23, ___, 29, 32

    A. 25
    B. 24
    C. 27
    D. 26

11. 20, 80, ___, 1280

    A. 240
    B. 320
    C. 640
    D. 960

12. 35, 45, 55, ___, 75

    A. 65
    B. 60
    C. 70
    D. 63

13. 9, 8, ___, 6, 5, 4

    A. 1
    B. 11
    C. 7
    D. 3

14. 23, 33, 43, ___, 63

    A. 13
    B. 53
    C. 73
    D. 50

15. 1, 1, ___, 1, 1, 1

    A. 1
    B. 3
    C. 2
    D. 4

Find the missing term.

16. $27 \div \triangle = 9$

    A. 4
    B. 3
    C. 12
    D. 18

17. $16 \times 4 = \square$

    A. 64
    B. 20
    C. 60
    D. 12

18. $444 + \triangle = 777$

    A. 227
    B. 317
    C. 222
    D. 333

19. $\square - 24 = 112$

    A. 88
    B. 136
    C. 96
    D. 126

20. $32 \div \triangle = 4$

    A. 16
    B. 12
    C. 9
    D. 8

21. $80 \times \square = 640$

    A. 8
    B. 80
    C. 40
    D. 20

22. $732 + \triangle = 900$

    A. 1632
    B. 178
    C. 168
    D. 268

23. $528 - 63 = \square$

    A. 465
    B. 591
    C. 456
    D. 1131

24. $126 \div 14 = \triangle$

    A. 9
    B. 11
    C. 12
    D. 14

25. $129 + \square = 200$

    A. 81
    B. 61
    C. 71
    D. 101

26. $17 \times \triangle = 51$

    A. 34
    B. 3
    C. 7
    D. 68

27. $533 - 441 = \square$

    A. 974
    B. 103
    C. 213
    D. 92

Chapter 7
Measurement

This chapter covers the following Georgia Performance Standards:

M4M	Measurement	M4M1.a, b, c

7.1   Measuring Weight Using Standard Units

In the United States, we use the Standard measuring system. The **standard system** uses ounces, pounds, and tons to measure weight. The chart below explains how the standard units of measure compare with one another.

Weight Measurement in the Standard System

Standard system measures in ounces, pounds, and tons
Abbreviations: ounce = oz; pound = lb.; ton = T
Equivalents: 16 ounces = 1 pound; 2,000 pounds = 1 ton

Ounces are the standard measurement used for small or very light items such as letters, spices, gold, beads, and candy.

Pounds are used to measure medium items such as packages, fruits, vegetables, cats, dogs, and people.

Tons are used to measure very large items such as cars and trucks, elephants and rhinos, things sent by railroad, or shipping containers.

Fill in the blanks in the following chart.

Item	Weight in oz.	Weight in lb.	Weight in tons
1 bottle of popcorn seeds	16 oz		0.0005 tons
2 packages of butter	32 oz		0.001 tons
1 medium elephant		4,000 lb.	2 tons
1 small car		1,000 lb.	0.5 ton
2 bags of candy	24 oz		0.00075 tons
1 railroad car full of cows		6,000 lb.	3 tons
1 full grown cat	112 oz		0.0035 tons
1 large pickup truck	32,000 oz	2,000 lb.	
1 large bottle of cinnamon	8 oz		0.00025 tons

Which unit of standard measurement would you use to measure the following?

1. The weight of a train car full of refrigerators would be measured in _____.

2. The weight of a package of feathers that could fill a teacup would be measured in _____.

3. The weight of a baby girl would be measured in _____ and _____.

4. The weight of a Labrador retriever would be measured in _____ and _____.

5. The weight of an airplane would be measured in _____.

6. The weight of newborn bluebird would be measured in _____.

7. The weight of a box of 24 hammers would be measured in _____.

8. The weight of a small candy bar would be measured in _____.

7.2   Measuring Mass Using Metric Measurements

Most of the world uses the Metric system. Grams and kilograms are the metric untis used to measure mass. Mass is the amount of matter packed into a space. The chart below explains how the metric units of measure for mass compare with one another.

Mass Measurement in the Metric System
Metric system measures in grams and kilograms
Abbreviations: gram = g; kilogram = kg
Equivalents: 1,000 grams = 1 kilogram

Grams are the metric measurement for very light mass items. A large paper clip has the mass of about one gram. Some things measured in grams are spices, gold, candy, letters, and just about any small item.

Kilograms are the metric measurement for medium and large items such as a person's mass, fruits and vegetables, cats and dogs, and larger items such as cars and trucks, large animals, things sent by railroad, or shipping containers. A kilogram is about two pounds.

Note: Many people confuse weight and mass. Weight measures the force of gravity pulling on an object. Mass is how tightly packed or dense matter is. The mass of an object is the same on Earth or on the moon because it does not depend on the force of gravity.

Fill in the blanks in the following chart.

Item	Mass in grams	Mass in kilograms
2 packages of butter	1,000 g	
1 full grown cat		3.5 kg
1 large container of beads		0.5 kg
1 bag of apples	2,000 g	
1 small car		500 kg
1 full grown chicken	2,500 g	
4 Labrador puppies	6,000 g	
1 box of brown sugar		1 kg
1 big box of stuffed animals	10,000 g	
1 bag of candy	250 g	

Which unit of metric measurement would you use to measure the following?

1. The mass of an adult man would be measured in _____.

2. The mass of a hummingbird would be measured in_____.

3. The mass of a button would be measured in _____.

4. The mass of a grizzly bear would be measured in _____.

5. The mass of a piece of gum would be measured in _____.

6. The mass of a basket of laundry would be measured in _____.

7. The mass of a bicycle would be measured in _____.

8. The mass of a pair of gold earrings would be measured in _____.

Chapter 7 Review

Which unit of standard measurement would you use to measure the following?

1. A bag of feathers _____

2. A railroad boxcar full of TVs

   _____

3. A hippopotamus _____

4. A single serving of candies _____

5. A bag of oranges _____

6. A tractor trailer full of hogs _____

7. A silver bracelet _____

8. A backpack full of books _____

Answer the following.

9. How many ounces are there in one pound? _____

10. How many pounds are there in one ton? _____

11. How many ounces are there in 3 pounds? _____

12. How many tons are there in 8,000 pounds? _____

13. How many pounds are there in 32 ounces? _____

1 TON

Which unit of metric measurement would you use to measure the following?

14. A gold ring _____

15. A teachers desk _____

16. A moose _____

17. A bag of baseballs _____

18. One rose _____

19. A railroad car full of shoes _____

20. A pair of socks _____

21. A case of bottled water _____

Answer the following.

22. How many grams are there in 1 kilogram? _____

23. How many grams are there in 4 kilograms? _____

24. How many kilograms are there in 7,000 grams? _____

Chapter 7 Test

Choose the correct response.

1. The weight of an elephant is best measured by which of the following?

   A. ounces

   B. grams

   C. tons

   D. pounds

2. The mass of a bumble bee is best measured by which of the following?

   A. kilograms

   B. grams

   C. tons

   D. pounds

3. How many ounces are there in one pound?

   A. 16

   B. 32

   C. 18

   D. 30

4. How many grams are in one kilogram?

   A. 500

   B. 1,000

   C. 1,500

   D. 2,000

5. How many pounds are in $\frac{1}{2}$ ton?

   A. 500

   B. 1,000

   C. 1,500

   D. 2,000

6. The weight of a battleship would be best measured by which of the following?

   A. ounces

   B. grams

   C. pounds

   D. tons

7. The weight of a large bag of rice would be best measured by which of the following?

   A. ounces

   B. grams

   C. pounds

   D. tons

8. How many tons are equal to 10,000 pounds?

   A. 5

   B. 4

   C. 2

   D. 1

9. How many pounds are equal to 32 ounces?

   A. 5

   B. 4

   C. 2

   D. 1

10. The mass of one piece of candy would be best measured by which of the following?

   A. kilograms

   B. grams

   C. pounds

   D. tons

11. How many grams are in five kilograms?

    A. 500

    B. 10,000

    C. 5,000

    D. 1,000

12. Bobby has a cat named Bubba that weighs 8 pounds. How much does Bubba weigh in ounces?

    A. 16

    B. 98

    C. 96

    D. 128

13. A shipping container full of microwaves weighs in at 4,000 pounds. How many tons is 4,000 pounds?

    A. 1

    B. 2

    C. 4

    D. 5

14. Ellen is taking her all her gold jewelry to the jeweler to find out its value. The jeweler will most likely weigh Ellen's jewelry by which measurement?

    A. ounces

    B. kilograms

    C. pounds

    D. tons

15. Which of the following is the best measurement for a carton of 12 blankets?

    A. tons

    B. grams

    C. ounces

    D. kilograms

Chapter 8
Angles and Rotation

This chapter covers the following Georgia Performance Standards:

M4M	Measurement	M4M2.a, b

8.1   Angle Names

Measuring angles is a necessary skill in many jobs such as carpentry and plumbing, sewing and quilting, architecture design, and graphic design. It is also used in the armed services, the space industry, and in many recreational skills such as archery, darts, and basketball.

We will begin by defining four kinds of angles.

A right angle measures 90°
and looks like the corner
of a page.

An acute angle is any angle
that measures less than 90°.

An obtuse angle is any angle
that measures more than
90°.

A straight angle measures
180° and is a straight
line with arrowheads on
both ends.

8.2 Angle Measurement

Angles are made up of two rays that look like arrows, with a common endpoint where the two rays meet. A **protractor** is used to measure angles. The protractor looks like a half of a circle, evenly divided into 180 degrees = 180° (° = degrees). To measure an angle using a protractor, the bottom of the angle is lined up on the bottom of the protractor, with the endpoint lined up in the middle of the protractor. The top part of the angle will point to the number of the degrees of the angle.

Example 1: Using the protractor below, we can see that $\angle FKA = 135°$.

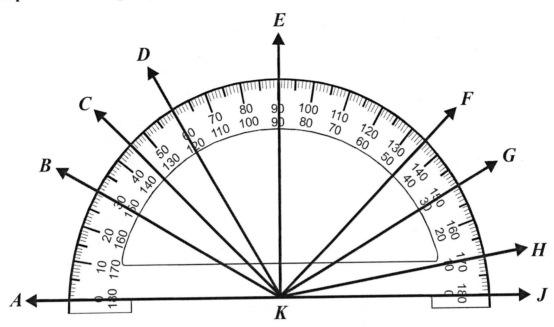

Hint: Notice that there are two rows of numbers on the protractor. One ray of the angle on the protractor always lines up with 0°. If the 0° is the inside number on the bottom of the protractor, measure the angle using the inside numbers. If 0° is the outside number on the bottom of the protractor, use the outside numbers to measure the angle.

Of the angles in the diagram above, find the measure of the angle and then tell the type of angle it is: right, acute, obtuse, or straight.

	Angle	Measure	Type of Angle
1.	$\angle BKJ$		
2.	$\angle HKJ$		
3.	$\angle CKJ$		
4.	$\angle GKJ$		
5.	$\angle DKJ$		
6.	$\angle EKJ$		
7.	$\angle AKJ$		

8.3   Other Ways To Measure Angles

If you don't have a protractor, you can still measure some angles accurately by folding a piece of paper. You already know that a corner of a piece of paper is a right angle and measures 90°. If you fold the paper meeting two sides together that share a corner, you will have one half of a 90° angle, which is a 45° angle. $90° \div 2 = 45°$. If you place another piece of paper next to this one you can add the two angles together $90° + 45° = 135°$.

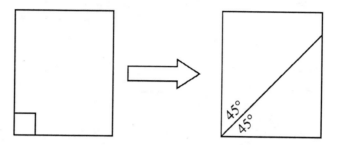

Try this yourself by folding the paper as stated above and then measure it with your protractor. See, it works!

Another tool used with angles is the **angle ruler**. The angle ruler looks like an ordinary ruler, however, the angle ruler pivots in the middle.

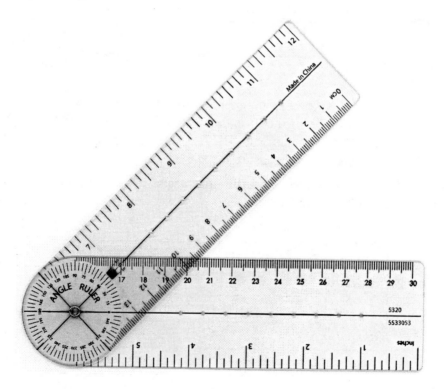

The movement ability allows you to not only measure angles but draw, or construct angles, too. The angle ruler can be made straight to be used as a 12" ruler.

8.4   Rotation

Rotation shows the movement of an object around a point. On a circle with degree markings, 360° is also the degree marking for 0°. That will be our starting point. There are 360° in a circle. If you twirl all the way around one time, you may say that you twirled around 360°. If you stop half way around, you may say that you twirled around 180°.

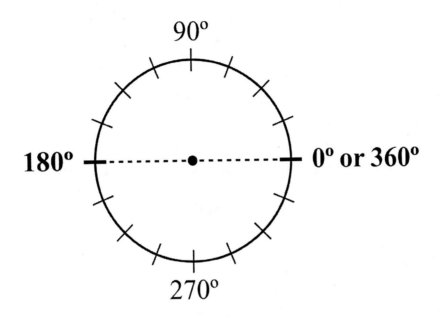

$$90^{o}$$

$$180^{o} \qquad 0^{o} \text{ or } 360^{o}$$

$$270^{o}$$

Example 2:

The chubby cheeked chipmunk has rotated 180°. It has twirled half way around.

Example 3:

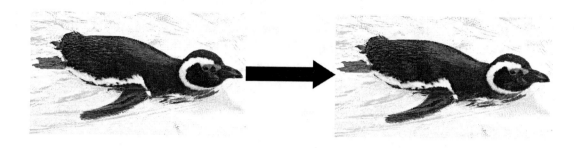

The penguin has turned 360°. The penguin has turned one full rotation. He is back where he started.

Examine the pictures below and determine if the object rotated 180° or 360°.

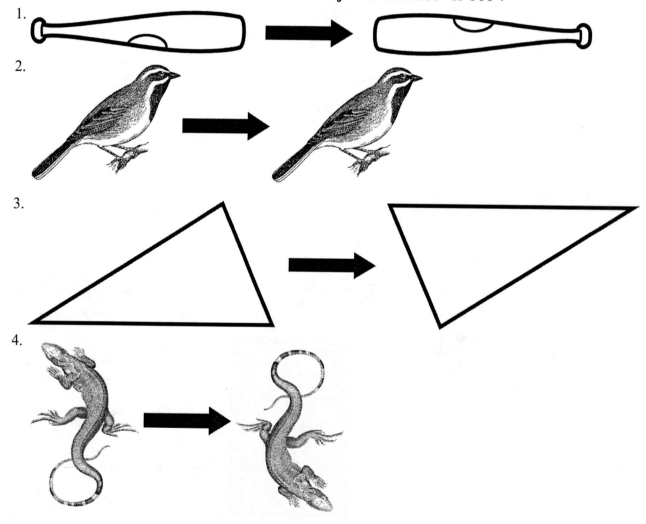

1.

2.

3.

4.

Chapter 8 Review

Using a protractor, measure the angles and then tell the type of angle it is: right, acute, obtuse, or straight.

1.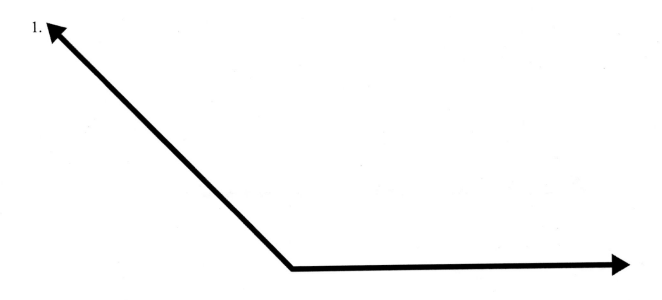

2.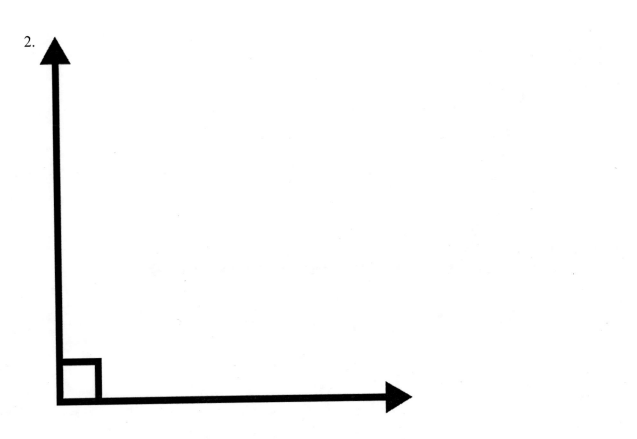

3.

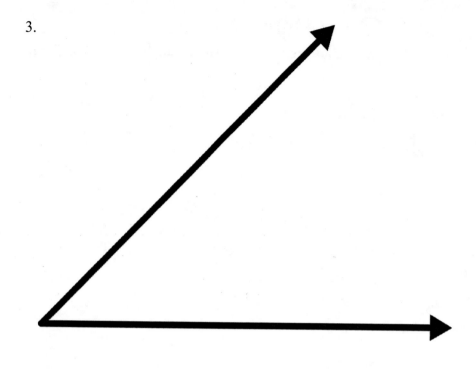

4.

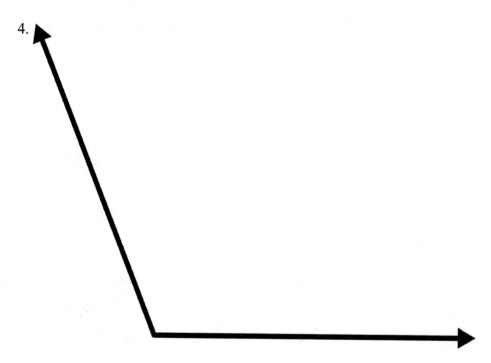

Examine the pictures below and determine if the object rotated 180° or 360°.

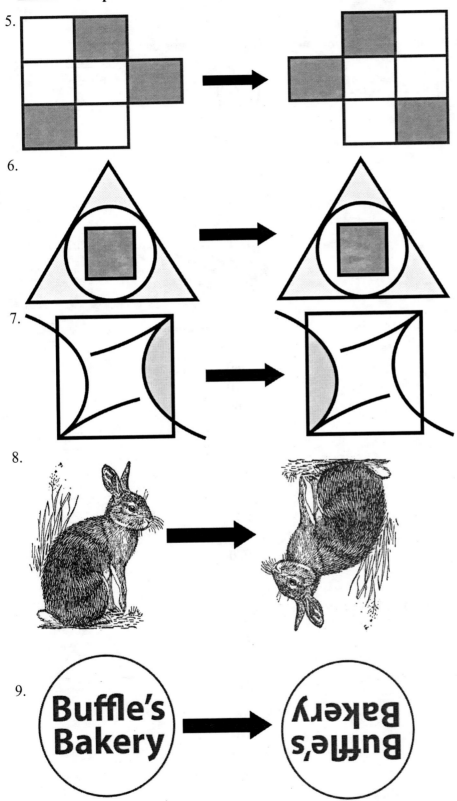

5.

6.

7.

8.

9. Buffle's Bakery

Chapter 8 Test

For questions 1–3, determine the type of angle.

1.

  A. right
  B. acute
  C. obtuse
  D. straight

2.

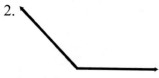

  A. right
  B. acute
  C. obtuse
  D. straight

3.

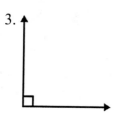

  A. right
  B. acute
  C. obtuse
  D. straight

4. How many degrees did the figure rotate?

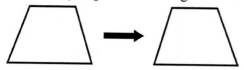

  A. 90°
  B. 180°
  C. 270°
  D. 360°

5. How many degrees did the figure rotate?

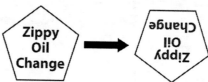

  A. 90°
  B. 180°
  C. 270°
  D. 360°

6. How many degrees did the figure rotate?

  A. 90°
  B. 180°
  C. 270°
  D. 360°

7. How many degrees did the figure rotate?

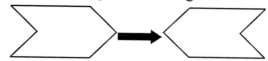

  A. 90°
  B. 180°
  C. 270°
  D. 360°

8. How many degrees did the figure rotate?

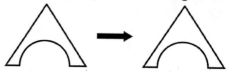

  A. 90°
  B. 180°
  C. 270°
  D. 360°

Chapter 9
Geometry

This chapter covers the following Georgia Performance Standards:

| M4M | Measurement | M4M2.c |
| M4G | Geometry | M4G1.a, b, c, d |

9.1   Identifying Triangles By Their Angles

right triangle
contains 1 right ∠

acute triangle
all angles are acute
(less than 90°)

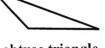

obtuse triangle
one angle is obtuse
(greater than 90°)

equilateral triangle
all three sides equal
all angles are 60°

scalene triangle
no sides equal
no angles equal

isosceles triangle
two sides equal
two angles equal

Facts about triangles:

Triangles have three sides.

Triangles have three angles.

The sum of the three interior angles of a triangle equals 180 degrees.

95

Identify the triangles below. Use the chart on the previous page to help you. Pay close attention to the definitions.

1.
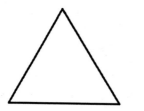

2.
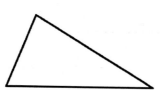

3.
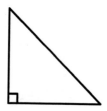

4.

5.
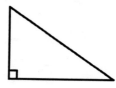

6.
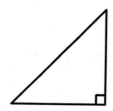

7.
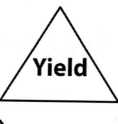

8.

Identify the measurement of the missing angle.

1.
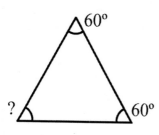

2.
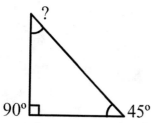

3.
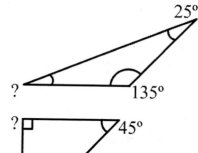

4.

9.2   Parallel and Perpendicular Lines

Parallel lines are two or more lines that are always equal distance from each other.

Example 1:   The two sides of a sheet of paper are parallel. The top and bottom of a sheet of paper are parallel.

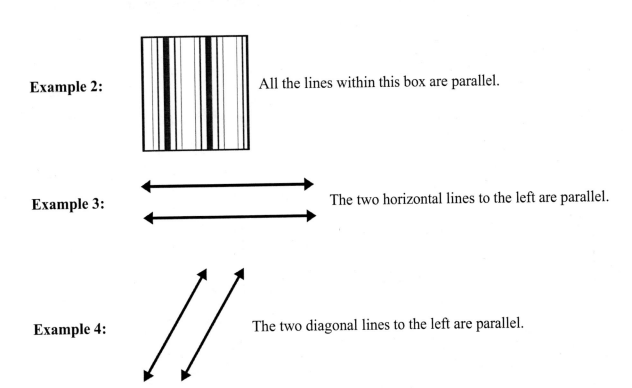

Example 2:   All the lines within this box are parallel.

Example 3:   The two horizontal lines to the left are parallel.

Example 4:   The two diagonal lines to the left are parallel.

Perpendicular lines are two lines that meet or cross each other at a right angle (90°).

Example 5:   Two lines that meet at a **right angle** are **perpendicular** to each other.

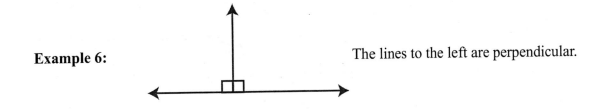

Example 6:   The lines to the left are perpendicular.

Example 7:

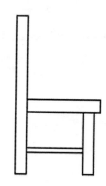

The back of the chair in the picture is perpendicular to the seat of the chair in the picture.
Also, the brace near the bottom of the chair is perpendicular to the back of the chair and to the front legs of the chair.
The seat of the chair is also perpendicular to the front legs of the chair.

Below are sets of lines and figures. State if they contain perpendicular or parallel lines, or both.

1.

4.

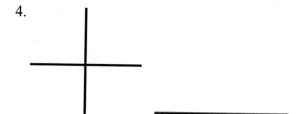

2.

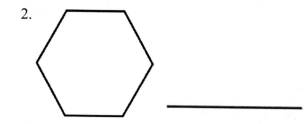

5.

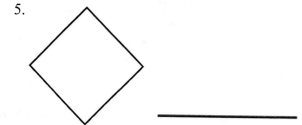

3.

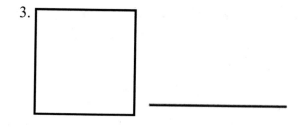

6.

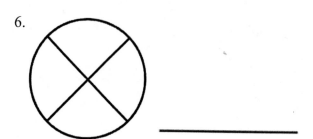

9.3 Quadrilaterals

A quadrilateral is a shape with 4 sides and 4 angles. The following chart describes types of quadrilateral.

Shape	Type of Quadrilateral	Type of Sides	Number of Sides	Angles
	General	4 straight lines	4	4
	Square	opposite sides are parallel	4 equal sides	4 right angles
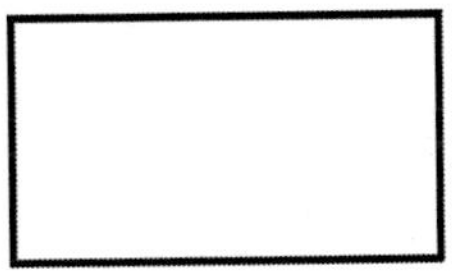	Rectangle	opposite sides are parallel	4; opposite sides are equal	4 right angles
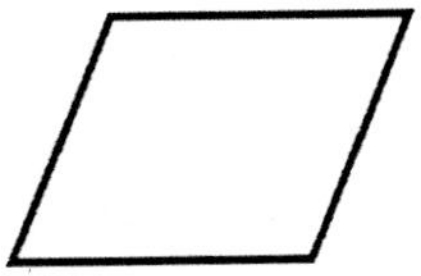	Parallelogram	opposite sides are parallel	4; opposite sides are equal	4
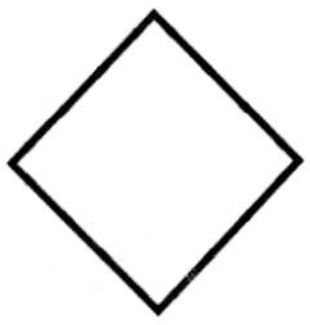	Rhombus	opposite sides are parallel	4 equal sides	4 right angles
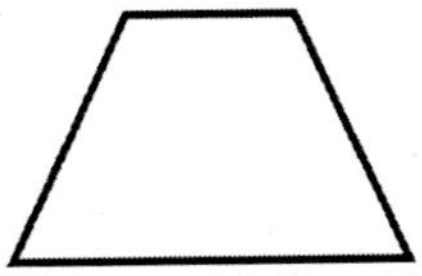	Trapezoid	1 pair of parallel sides	4	4

If you study the chart above carefully, you will see that some of the definitions of one quadrilateral may fit other quadrilaterals.

Example 8:   A parallelogram is defined as a four-sided figure with 2 pairs of parallel sides that are opposite each other, and 2 pairs of equal sides. A square, a rectangle, and a rhombus also fit that description. Therefore, we may say that squares, rectangles, and rhombi are also parallelograms.

Example 9:   A rectangle is defined as a four-sided figure with 2 pairs of parallel sides that are opposite each other, 4 right angles, and 2 pairs of equal sides. A square also fits that description. While a square can be a rectangle, a rectangle cannot be a square because a square by definition must have 4 equal sides.

Identify the following shapes by their quadrilateral name.

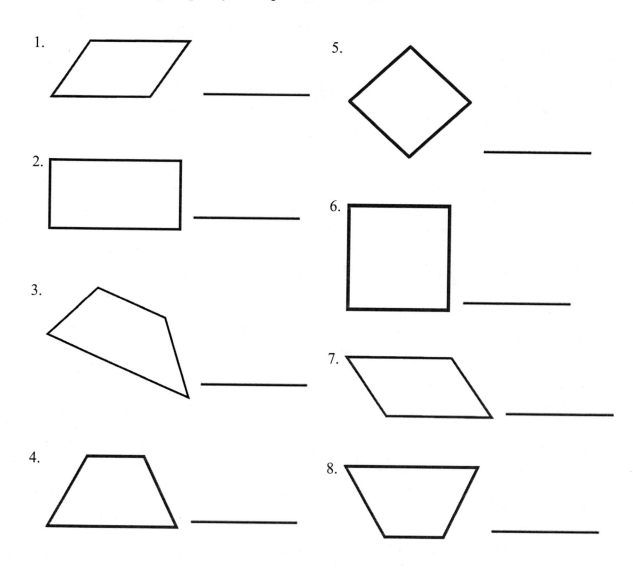

1. _____

2. _____

3. _____

4. _____

5. _____

6. _____

7. _____

8. _____

Chapter 9 Review

Identify the triangles below.

1. 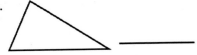 _____

2. 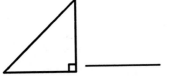 _____

3. 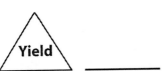 _____

4.  _____

5.  _____

6. 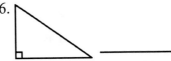 _____

Below are figures. State if the figure contains perpendicular lines, parallel lines, or both types of lines.

7.  _____

8.  _____

9. 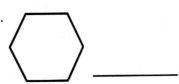 _____

10.  _____

Identify the following shapes by their quadrilateral names.

11.  _____

12. 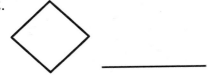 _____

13. 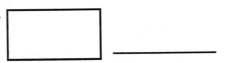 _____

14.  _____

15. True or false? An isosceles triangle has three equal sides and angles. _____

16. True or false? A square is a rectangle. _____

Chapter 9 Test

1. Identify the figure below.

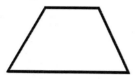

   A. The figure above is a rhombus.
   B. The figure above is a trapezoid.
   C. The figure above is a scalene triangle.
   D. The figure above is a rectangle.

2. An equilateral triangle is defined as having:

   A. all 3 sides and 3 angles are equal
   B. no sides or angles are equal
   C. 2 sides and 2 angles are equal
   D. 1 angle is a right angle (90°)

3. True or false? - Parallel lines are equally distant from each other.

   A. true
   B. false

4. Identify the figure below.

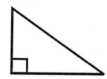

   A. The figure above is an acute triangle.
   B. The figure above is an obtuse triangle
   C. The figure above is a scalene triangle.
   D. The figure above is a right triangle

5. Choose the correct statement about the figure below.

   A. The figure above has 2 sets of parallel lines and 0 perpendicular lines
   B. The figure above has 2 sets of parallel lines and 4 perpendicular lines
   C. The figure above has 0 parallel lines and 2 perpendicular lines
   D. The figure above has 1 set of parallel line and 1 perpendicular line

6. Choose the correct statement about the figure below.

   A. The figure above has 2 parallel lines and 0 perpendicular lines
   B. The figure above has 0 parallel lines and 0 perpendicular lines
   C. The figure above has 0 parallel lines and 2 perpendicular lines
   D. The figure above has 1 parallel line and 1 perpendicular line

7. Identify the figure below.

   A. The figure above is a scalene triangle
   B. The figure above an equilateral triangle
   C. The figure above is an isosceles triangle
   D. The figure above is a right triangle

8. True or false? - A square is a rectangle, but a rectangle is not a square.

   A. true

   B. false

9. Define a parallelogram.

   A. A parallelogram has 3 equal angles and sides

   B. A parallelogram has 2 pairs of parallel lines opposite each other and 4 equal sides

   C. A parallelogram has 2 pairs of parallel lines opposite each other and 2 pairs of equal sides

   D. A parallelogram has 3 pairs of parallel lines beside each other and 2 equal angles

10. Define an acute triangle.

   A. An acute triangle has 3 angles that are less than 90° each.

   B. An acute triangle has 2 angles and sides the same.

   C. An acute triangle has 4 sides and 4 angles the same.

   D. An acute triangle has at least 1 angle over 90°.

11. What is the proper name for the quadrilateral below?

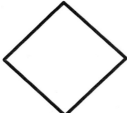

   A. The figure above is a rectangle.

   B. The figure above is a baseball diamond.

   C. The figure above is a trapezoid.

   D. The figure above is a rhombus.

12. What is the name of the triangle in the middle of the figure below?

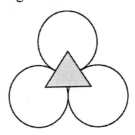

   A. The figure above contains an equilateral triangle.

   B. The figure above contains a rhombus triangle.

   C. The figure above contains a right triangle.

   D. The figure above contains an obtuse triangle.

13. Which statement is true about the figure below?

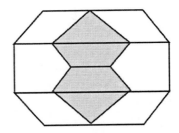

   A. The figure above contains circles, squares, and triangles.

   B. The figure above contains triangles, a rectangle, and trapezoids.

   C. The figure above contains squares, rectangles, and triangles.

   D. The figure above contains rectangles, a rhombus, and triangles.

14. True or false? - An obtuse triangle has 1 angle greater than 90°.

   A. true   B. false

Chapter 10
Solid Geometry

This chapter covers the following Georgia Performance Standards:

| M4G | Geometry | M4G2.a, b, c |

10.1 Cubes and Rectangular Prisms

Cube   Rectangular Prism

Vertex

Face or side

Edge

Face or side

Edge

	Cube	**Rectangular Prism**
Number of sides	6	6
Number of faces	6	6
Number of vertices	8	8
Number of edges	12	12
Sides	all the same size	size can vary

Each face (side) of a cube or rectangular prism has two sets of parallel lines. Each face has 4 right angles where the sides of the face meet.

All 6 faces of a cube or rectangular prism meet together and are perpendicular to each other.

Example 1: If a cube was opened up and laid flat, it would look like the figure below. Note how each face, or side, of a cube is a square.

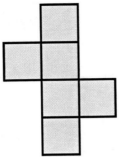

Example 2: If a rectangular prism were opened up and laid flat, it would like the figures below. Note how 4 of the faces, or sides of a rectangular prism looks like a rectangle, and 2 sides that can look like squares (figure 1) or rectangles (figure 2). Like the cube, all 6 sides have 2 sets of parallel lines meeting at 4 right angles.

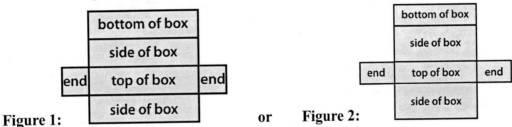

Figure 1:   or   **Figure 2:**

Below is a list of many objects. State if the object described is shaped like a cube or a rectangular prism.

1. A shoebox —————————

2. A toy block —————————

3. A 2 inch thick book, 6 inches by 9 inches —————————

4. A 3 inch tall stack of 3 inch square crackers —————————

5. A 2 foot long carton, 1 foot high, of oranges —————————

6. A box of cereal —————————

7. A stack 4 feet high, of 4 feet square rugs —————————

10.2   Solid Geometric Figures

Cylinder   Cube   Rectangular Prism   Cone   Sphere   Pyramid

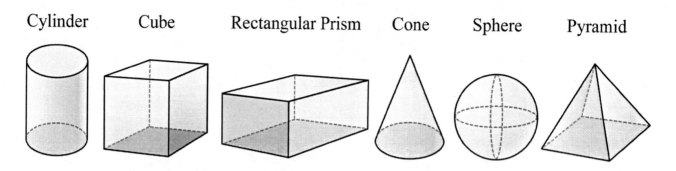

See the chart below for the definitions and parts of the geometric solids pictured above. If you would like to construct models for these figures, you can cut out the shapes from poster board and tape them together to hold the shape upright. You may need to support the construction on the inside with some sort of "stuffing" such as crumpled paper if your poster board isn't stiff enough.

Figure	Description	Shapes that make up the figure	Objects that use the shape
Cylinder	2 circular faces, connected by a rectangular middle		Cans, tubes, jars
Cube	6 square faces, all the same size		Boxes, containers
Rectangular Prism	6 faces, 2 faces are the same size and the other 4 faces are the same size		Boxes, books, containers
Cone	Circular face, fan shaped piece connected to circular bottom, coming to a vertex at the top		A waffle cone for ice cream, a party hat
Square Pyramid	Square on the bottom, triangles attached to each side and meeting at a vertex.		Egyptian pyramids

Look at the solid geometric figures below and state the name of the figure.

1. 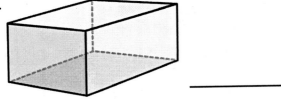 _____

4. 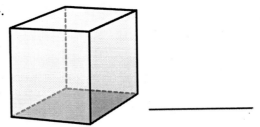 _____

2. 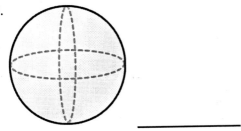 _____

5. 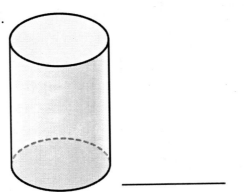 _____

3. 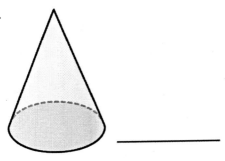 _____

6. 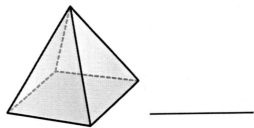 _____

Identify the solid given from the descriptions below.

7. A solid of 6 equal square faces _____

8. A solid of 2 circles and 1 rectangle _____

9. A solid of 1 square and 4 triangles _____

10. A solid of 2 squares and 4 rectangles _____

Chapter 10 Review

Identify the solid geometric solids in the pictures below.

1.
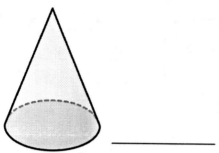
_____

4.
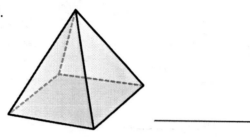
_____

2.
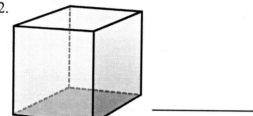
_____

5.
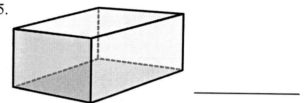
_____

3.
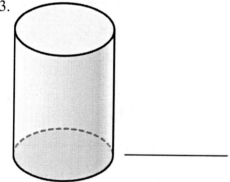
_____

6.
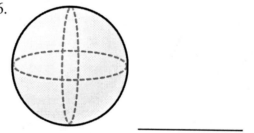
_____

Below is a list of objects. State which geometric solid the object is most shaped like.

7. A can of soda _____

8. A basketball _____

9. A square box _____

10. A shoebox _____

11. An Egyptian pyramid _____

12. A baseball _____

Chapter 10 Test

Choose the best answer for the following questions.

1. What will the figure below be when it is put together?

   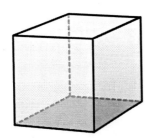

   A. The figure above is a rectangular prism.
   B. The figure above is a cube.
   C. The figure above is a pyramid.
   D. The figure above is a cylinder.

2. What solid figure is made up of 1 square and 4 triangles?

   A. rectangular prism
   B. cylinder
   C. square pyramid
   D. sphere

3. Which object below resembles a cylinder?

   A. shoebox
   B. can of vegetables
   C. basketball
   D. book

4. The figure below is made up of what parts?

   A. 2 squares and 4 rectangles
   B. 2 circles and 1 rectangle
   C. 1 square and 4 triangles
   D. 6 equal squares

5. Which object below resembles a sphere?

   A. Egyptian pyramid
   B. the planet Earth
   C. a 6 inch tube
   D. a football

6. Which figure is made of the parts shown below?

   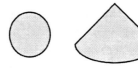

   A. square pyramid
   B. cylinder
   C. rectangular prism
   D. cone

7. Which figure has 6 faces all the same size, 8 vertices, and 12 edges?

   A. cube
   B. rectangular prism
   C. cylinder
   D. square pyramid

8. Which figure is made of the parts shown below?

   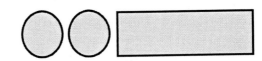

   A. cone
   B. rectangular prism
   C. cylinder
   D. cube

9. If you glued 2 cubes together side by side, what new figure would you have?

   A. square pyramid
   B. rectangular prism
   C. cone
   D. cube

10. Which list of items all resemble the figure below?

   A. a soda can, a box of raisins, and a baseball
   B. a soda can, a can of green beans, and a jar of peanut butter
   C. a soda can, a box of chocolate, and a football
   D. a soda can, a shoebox, and a jar of peanut butter

11. Which list of items all resemble a sphere?

   A. a basketball, a globe, and a shoebox
   B. a basketball, a jar of jelly, and a football
   C. the planet Earth, a soccer ball, and an orange
   D. the planet Earth, a basketball, and a football

12. Which figure is made up of the parts below?

   A. sphere
   B. cone
   C. cube
   D. square pyramid

Chapter 11
Coordinate Systems

This chapter covers the following Georgia Performance Standards:

M4G	Geometry	M4G3.a, b, c

11.1   Defining a Coordinate System and Applying Ordered Pairs

A **number line** allows you to graph points with one value. For instance, the number 2 is graphed on a number line as shown below.

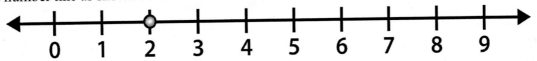

A **coordinate system** allows you to graph points with two values. A coordinate system has two number lines, the horizontal line called the x-axis and the vertical line called the y-axis. The point where the x and y axes intersect is called the **origin**. Each point on the coordinate system is designated by an **ordered pair** of coordinates. The ordered pairs are enclosed in parentheses, with the x-axis named first and the y-axis named second: (x, y). These values are called **coordinates**. An easy way to remember which number comes first is the axes are in alphabetical order. X comes before Y in the alphabet, and horizontal comes before vertical in the alphabet.

To identify a coordinate pair, simply count from the origin, $(0, 0)$.

Example 1:   The ordered pair in the coordinate system below is $(2, 3)$.

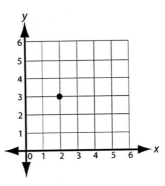

The point is above the number 2 on the x-axis, and the point is across from the 3 on the y-axis. Therefore, the coordinate pair is $(2, 3)$.

Example 2:   The ordered pair in the coordinate system below is $(1, 4)$.

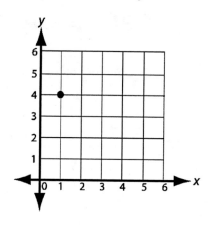

The point is above the number 1 on the x-axis and the point is across from the number 4 on the y-axis. Therefore, the coordinate pair is $(1, 4)$.

If the ordered pair has a zero in it, the marker will land on either the horizontal or vertical number line. If the ordered pair begins with a zero, such as $(0, 5)$, the marker for the ordered pair will land on the vertical number line. If the ordered pair has a zero in the second place, such as $(8, 0)$, the marker for the ordered pair will land on the horizontal number line.

Locate the point on the coordinate systems below and name the ordered pair.

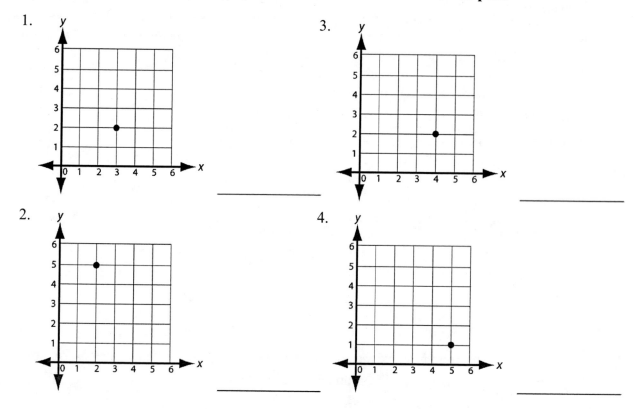

1.

2.

3.

4.

11.2 Graphing Points on a Coordinate Plane

When given an ordered pair to graph, find the number on the x-axis first, then count up to the number on the y-axis.

Example 3: Find the point where the ordered pair $(1,3)$ would be placed on the coordinate plane below.

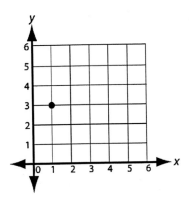

The number in the ordered pair for the x-axis is 1. So first, go to the number 1 on the x-axis. The number in the ordered pair for the y-axis is 3, so count up from the 1 to the third line. This is where you place the point for the ordered pair $(1,3)$.

Graph the following ordered pairs.

1. $(3,5)$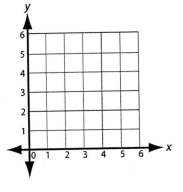

3. $(2,2)$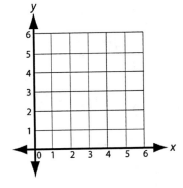

2. $(4,0)$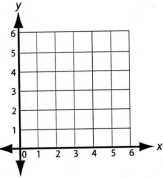

4. $(0,3)$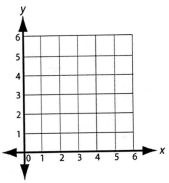

5. $(5, 4)$

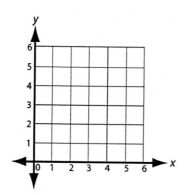

9. $(0, 1)$

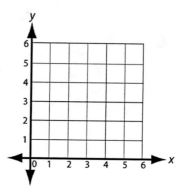

6. $(1, 6)$

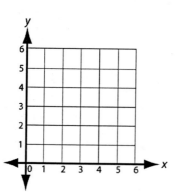

10. $(6, 4)$

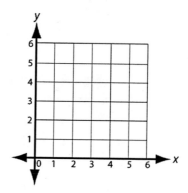

7. $(2, 3)$

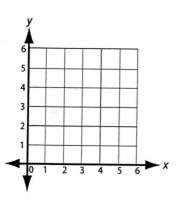

11. $(4, 5)$

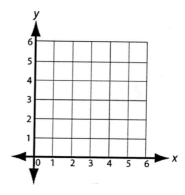

8. $(3, 1)$

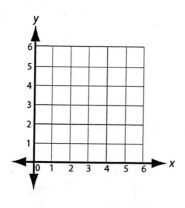

12. $(3, 0)$

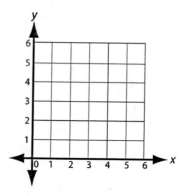

Chapter 11 Review

Locate the point in the coordinate systems below, and name the ordered pair.

1.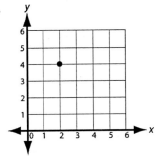

_____

2.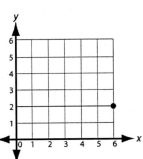

_____

3.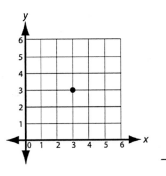

_____

4.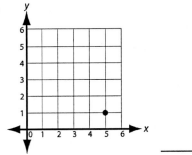

_____

Graph the following ordered pairs.

5. $(6, 6)$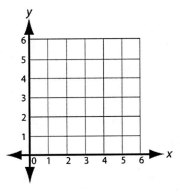

7. $(5, 2)$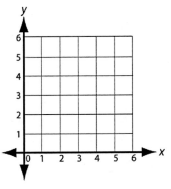

6. $(0, 4)$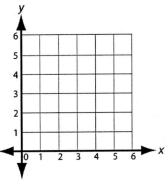

8. $(4, 3)$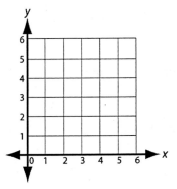

Chapter 11 Test

1. Which statement is true?

   A. The x-axis is the vertical axis.
   B. The y-axis is the vertical axis.
   C. The x-axis is parallel to the y-axis.
   D. The y-axis is the horizontal axis.

2. Locate the point in the coordinate system below and name the ordered pair.

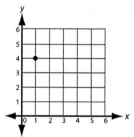

   A. $(1, 0)$
   B. $(0, 4)$
   C. $(1, 4)$
   D. $(4, 1)$

3. In the ordered pair $(8, 3)$, which number is the y coordinate?

   A. 8
   B. 0
   C. 11
   D. 3

4. Locate the point in the coordinate system below and name the ordered pair.

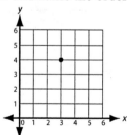

   A. $(3, 0)$
   B. $(0, 4)$
   C. $(3, 4)$
   D. $(4, 3)$

5. In the ordered pair $(7, 2)$, which number is the x-axis?

   A. 7
   B. 0
   C. 2
   D. 9

6. Locate the point in the coordinate system below and name the ordered pair.

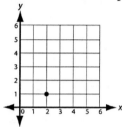

   A. $(1, 2)$
   B. $(0, 2)$
   C. $(1, 0)$
   D. $(2, 1)$

7. Which statement is true?

   A. The y-axis is the horizontal axis.
   B. The y-axis runs parallel to the x-axis.
   C. The x-axis is the horizontal axis.
   D. The x-axis is the vertical axis.

8. In the ordered pair $(3, 5)$, which number is the y coordinate?

   A. 5
   B. 3
   C. 8
   D. 2

Locate the point in the coordinate systems below, and name the ordered pairs.

9.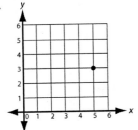

A. $(5, 0)$
B. $(0, 3)$
C. $(5, 3)$
D. $(3, 5)$

12.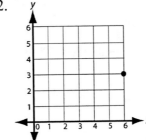

A. $(6, 0)$
B. $(0, 3)$
C. $(3, 6)$
D. $(6, 3)$

10.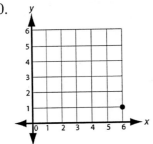

A. $(6, 0)$
B. $(6, 1)$
C. $(1, 6)$
D. $(0, 1)$

13.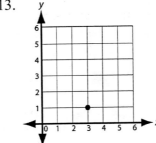

A. $(3, 0)$
B. $(3, 1)$
C. $(1, 3)$
D. $(0, 1)$

11.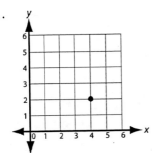

A. $(4, 0)$
B. $(2, 4)$
C. $(4, 2)$
D. $(0, 4)$

14.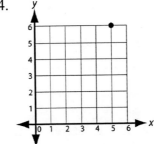

A. $(5, 6)$
B. $(6, 5)$
C. $(5, 0)$
D. $(0, 6)$

Chapter 12
Data Analysis

This chapter covers the following Georgia Performance Standards:

M4D	Data Analysis	M4D1.a, b, c, d

12.1   Bar Graphs, Line Graphs, and Pictographs

Graphs are used to display data. When using graphs, you can see how the data are compared very quickly. Some types of graphs work best with certain data. Some data can be shown in different types of graphs. The average temperature of seasons is one example. Notice how the data in all three graphs are the same, however, the graphs look different.

Average Temperature of Seasons

Temperature	Season
70°	spring
95°	summer
75°	fall
60°	winter

Average Temperature of Seasons

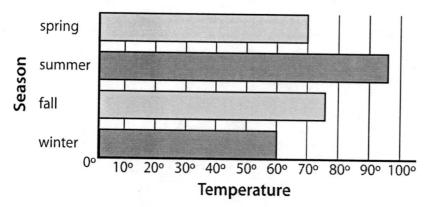

This is a bar graph. The bars on a bar graph may be shown horizontally or vertically.

Average Temperature of Seasons

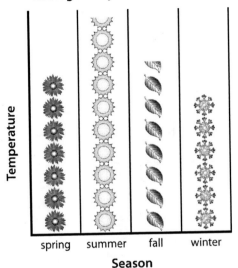

This is a line graph.
Line graphs show
change over a period
of time.

Average Temperature of Seasons

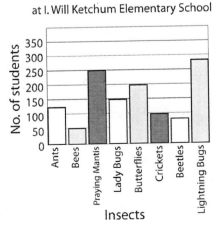

This is a pictograph. Pictographs use
symbols to represent certain amounts.
If half of a symbol is shown, it means
half of the given amount is represented.
Pictographs always have a key to show
the amount each symbol represents.
Each symbol represents 10°.

Answer the questions for the graph below and for each graph on the next page.

Favorite Insects of 4th Graders
at I. Will Ketchum Elementary School

1. How many students said ants were their favorite insects?

2. How many students said lady bugs were their favorite?

3. What is the total of the students who preferred crickets and beetles?

4. Which insect was the favorite of the most students?

Answer the questions for each graph.

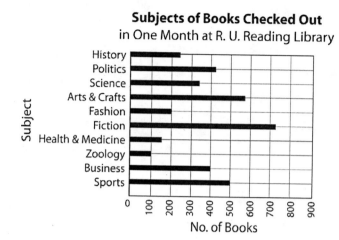

5. How many arts & crafts books were checked out?

6. How many fiction books were checked out?

7. What is the total number of fashion books and sports books checked out?

8. How many science books were checked out?

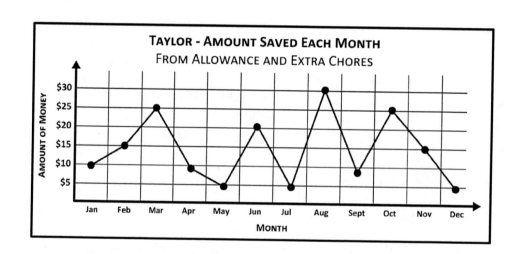

9. How much money did Taylor save in March?

10. How much money did Taylor save in June and July?

11. How much more money did Taylor save in August compared to July?

12. How much money did Taylor save in December?

12.2   Making Graphs

You can make a graph out of almost any set of figures.

Billy's Bass Tournament was held in June and the following people participated in the contest for the biggest bass caught. Here is the list of participants and the size of their catch:

Name	Weight of Bass Caught
Bob Mangel	1 and 1/2 pounds
Todd Archer	4 and 1/2 pounds
Katie Albrite	3 pounds
Matt Cisco	5 and 1/2 pounds
Al Wilson	2 pounds
Rob Green	6 and 1/2 pounds
Herb Taylor	1 pound
Artie Bale	3 pounds
Sarah Lin	4 pounds

An easy way to compare everyone's catch is to graph the data. The only data you are given are the names of the participants and the size of the catch, so we will use the data we are given by putting the names along one side of the graph and possible weights along the bottom of the graph. See the completed graph below.

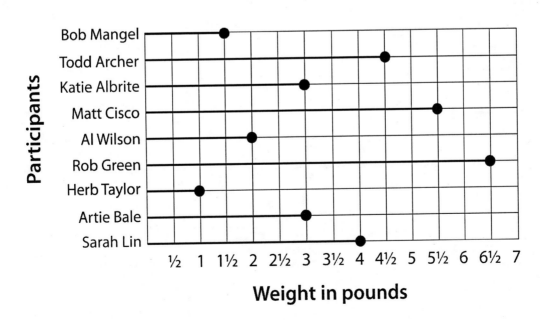

Make a bar graph out of the following set of data. Put the number of wins along one side of the graph and the names of the teams along the bottom of the graph. Make a bar graph to show the data. Be sure to label the side and the bottom to make it clear what data are shown.

Name of Little League Teams	Number of Wins
Bears	3
Cougars	2
Woodchucks	7
Eagles	5
Wolves	6
Gators	4

LITTLE LEAGUE TEAM WINS

Copyright © American Book Company

12.3   Problems With Data and Graphs

Sometimes a graph will show an error in the data set that may not have been found just by looking at the data. For instance, there may be repeats in a data set, or perhaps, missing information. A graph can make finding an error easy to identify.

Favorite Hard Candy of 100 Students Surveyed	
Type of Hard Candy	Number of Students
Peppermint	40
Butterscotch	25
Cinnamon	15
Spearmint	20
Peppermint	40

Notice at the top of the chart, it says there were 100 students surveyed. Something must be wrong, because the total number of students is 140 in the chart! Does this mean that 140 students were actually surveyed or is there a problem with the list? A careful check of the list shows that Peppermint candies are listed twice. If we remove the repeated Peppermint line of data, the total of the number of students comes to 100 – a perfect match. Now a graph of the data can be made.

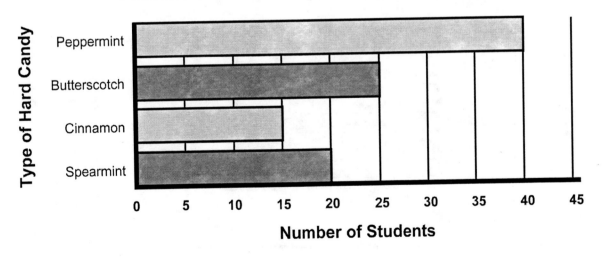

FAVORITE HARD CANDY OF **100** STUDENTS SURVEYED

The following graphs have problems with missing information or duplicate data items. State the problem each graph has.

1.

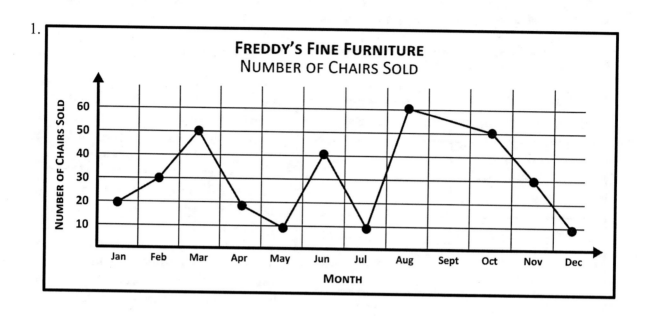

2.

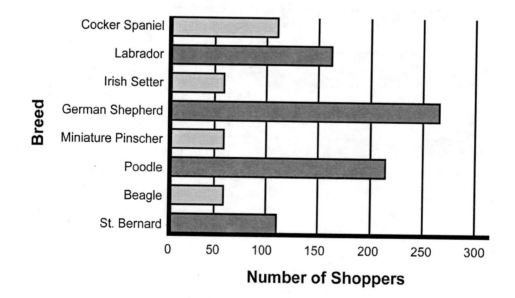

Copyright © American Book Company

Chapter 12 Review

Answer the questions for the graph below.

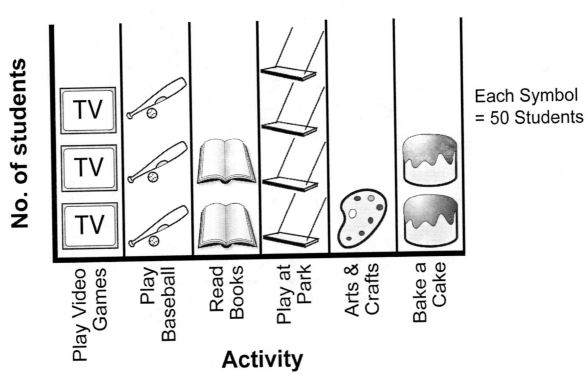

**What 4th Graders at
N. Joying Elementary School
Like to Do on Saturdays**

No. of students

Each Symbol
= 50 Students

Play Video Games | Play Baseball | Read Books | Play at Park | Arts & Crafts | Bake a Cake

Activity

1. How many students like to play baseball on Saturdays?

2. How many student prefer to do arts and crafts?

3. How many students enjoy playing at the park?

4. How many students prefer playing video games compared to those that prefer reading books?

Answer the questions for the graph below.

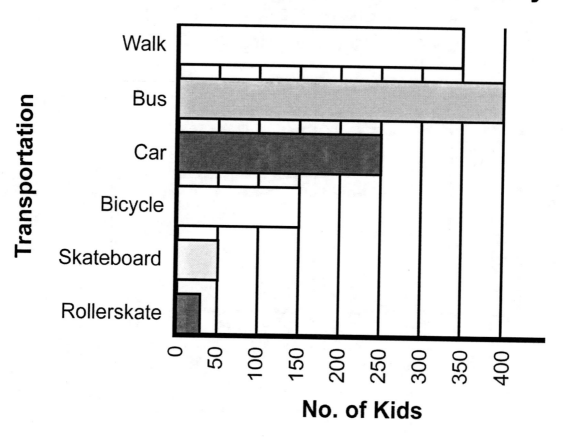

How Kids Get to School
at R. U. Walker Elementary

5. How many kids skateboard to school at R.U. Walker Elementary?

6. What is the total number of kids who ride a bicycle or rollerskate to school?

7. How many kids walk to school?

8. How many more kids ride the bus to school than ride in a car to school?

The following graph has a problem. State the problem for the graph.

9.

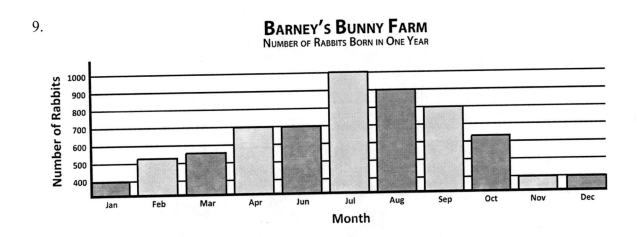

BARNEY'S BUNNY FARM
NUMBER OF RABBITS BORN IN ONE YEAR

Using the graph above, answer the following questions.

10. How many rabbits were born April? _____

11. How many rabbits were born in August and September? _____

12. Which month showed the greatest number of rabbits born? _____

13. Which 3 months showed the least number of rabbits born? _____

Chapter 12 Test

Using the graph below, answer the following questions.

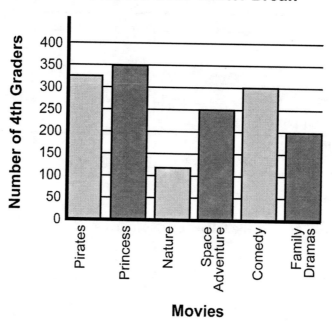

**Types of Movies Seen by
4th Graders Over Winter Break**

1. About how many 4th graders saw space adventure movies over winter break?

    A. 350
    B. 250
    C. 300
    D. 200

2. About how many 4th graders saw pirate and princess movies?

    A. 675
    B. 525
    C. 650
    D. 625

3. Which kind of movie did the least number of 4th graders watch?

    A. pirates
    B. family drama
    C. comedy
    D. nature

Using the graph below, answer the following questions.

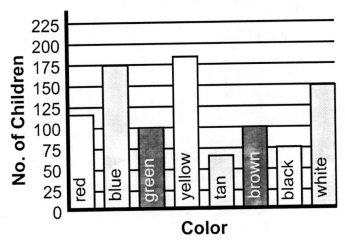

4. What kind of data does the graph above show?

  A. students' favorite colors
  B. the number of children attending school
  C. the color of shirts worn on the first day of school
  D. the favorite kind of apple among students

5. Which color of shirt was worn the most on the first day of school?

  A. yellow
  B. blue
  C. pink
  D. brown

6. What is the total number of green and white shirts shown in the graph?

  A. 225
  B. 350
  C. 250
  D. 325

Using the graph below, answer the following questions.

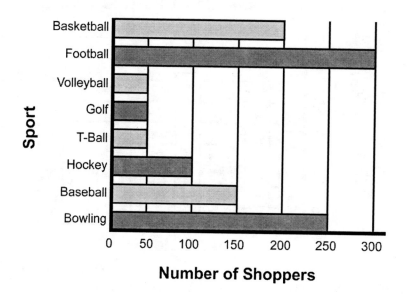

Survey of 1,000 Shoppers at Mega-Mall
Question: What is your favorite sport to watch in person?

Number of Shoppers

7. There is something wrong with the graph above. Which statement best describes what is wrong?

 A. Our family was not included in the survey.
 B. Ping Pong was not included among the favorite sports.
 C. The survey says 1,000 shoppers were surveyed, but the graph shows 1,050 shoppers were surveyed.
 D. The survey says 1,000 shoppers were surveyed, but the graph shows 1,150 shoppers were surveyed.

8. How many shoppers chose football as their favorite sport to watch?

 A. 200
 B. 250
 C. 275
 D. 300

9. How many more shoppers chose bowling than volleyball as their favorite sport to watch?

 A. 300
 B. 200
 C. 100
 D. 50

Using the graph below, answer the following questions.

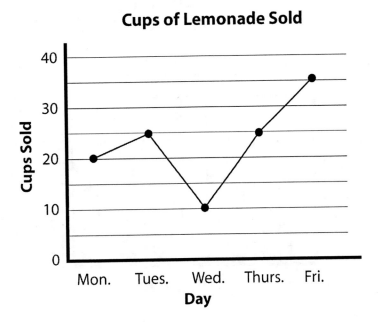

Cups of Lemonade Sold

10. The graph shows the cups of lemonade sold. How many cups of lemonade were sold on Thursday?

    A. 10
    B. 15
    C. 25
    D. 35

11. Which day were the most cups of lemonade sold?

    A. Friday
    B. Thursday
    C. Wednesday
    D. Tuesday

12. What is the total number of cups of lemonade sold in the week?

    A. 35
    B. 100
    C. 115
    D. 125

Using the graph below, answer the following questions.

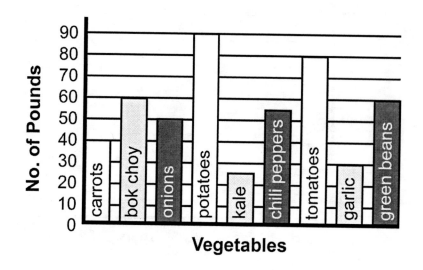

13. What is the total number of pounds of kale, garlic, and onions from the garden?

    A. 105 pounds
    B. 200 pounds
    C. 150 pounds
    D. 250 pounds

14. What is the total number of pounds of green beans from Mrs. Eatyore's vegetable garden?

    A. 65
    B. 720
    C. 40
    D. 60

15. How many more pounds of tomatoes are there than bok choy?

    A. 50
    B. 20
    C. 80
    D. 60

Practice Test 1

Part One

1. What is sixty eight thousand, two hundred twenty-eight in standard form?

    A. 28,628

    B. 68,228

    C. 68,668

    D. 28,268

M4N1b

2. Identify the correct operation.
    27 ☐ 3 = 9

    A. ÷

    B. ×

    C. +

    D. −

M4A1b

3. Marcie had a black lab puppy that always chased its tail two times before plopping down for a nap. How many degrees of rotation does Marcie's puppy make before each nap?

    A. 120°
    B. 360°
    C. 540°
    D. 720°

M4M2b

4. Round the following number to the nearest hundred.

    645,218

    A. 645,218
    B. 645,000
    C. 645,220
    D. 645,200

M4N2a

5. Karisha worked after school in a pet store. Her job included restocking the shelves of pet food. On Wednesday, Karisha put eight 40 pound bags and six 20 pound bags and seven 8 ounce cans of dog food on the shelves. What is the total weight of all the dog food Karisha put on the shelves on Wednesday?

    A. 440 pounds, 8 ounces

    B. 443 pounds, 12 ounces

    C. 443 pounds, 8 ounces

    D. 440 pounds, 12 ounces

M4M1b

6. Lila went shopping with her mom to buy things for Lila's birthday party. Lila chose 1 package each of balloons that cost $1.50, paper plates at $2.75, napkins at $1.90, and paper cups at $2.30. If Lila's mother pays for the items with one denomination of money, which bill is the closest one that Lila's mother should use?

    A. A five dollar bill

    B. A ten dollar bill

    C. A twenty dollar bill

    D. A fifty dollar bill

M4N2d

7. Which lines are perpendicular?

A.

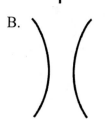

B.

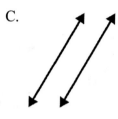

C.

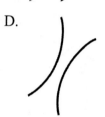

D.

M4G1b

8. Find the sum. $32.98 + 36.74$.

A. 60.72
B. 67.72
C. 69.72
D. 69.63

M4N5c

9. Find the product. $722 \times 12 =$

A. 8,664
B. 8,662
C. 8,466
D. 8,644

M4N3

10. Solve.
$$4 \times (3 + 2) + 2 \times (7 - 3) =$$

A. 40
B. 28
C. 12
D. 26

M4N7b

11. How many cups of lemonade were sold on Monday and Friday?

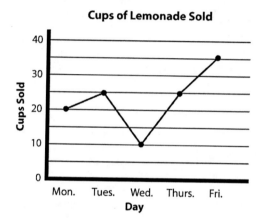

A. 20
B. 35
C. 50
D. 55

M4D1a

12. Round the following decimal to the nearest whole number.

65,841.36

A. 65,840
B. 65,800
C. 65,841
D. 65,841.4

M4N2c

134

13. Which ordered pair represents the point on the coordinate plane below?

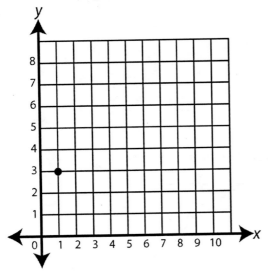

A. $(3, 1)$
B. $(3, 0)$
C. $(1, 3)$
D. $(1, 0)$

M4G3b

14. Put the following numbers in order from least to greatest.

$$9.2 \quad 4.9 \quad 9.4 \quad 2.9$$

A. 2.9   4.9   9.4   9.2
B. 2.9   4.9   9.2   9.4
C. 9.2   9.4   4.9   2.9
D. 9.4   9.2   4.9   2.9

M4N5b

15. $826 \div 4 = $ _____

A. 206 r2
B. 206 r4
C. 204 r2
D. 208 r2

M4N4b

16. What number in the following sentence is the divisor?

$$507,072 \div 1,112 = 456$$

A. 507,072
B. 1,122
C. 465
D. 1,112

M4N4c

17. What is the difference in weight between the bass caught by Rob Green and the bass caught by Katie Albrite?

BILLY'S BASS TOURNAMENT

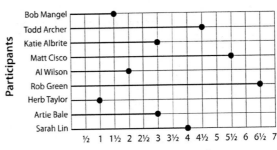

A. $2\frac{1}{2}$ pounds
B. 3 pounds
C. $3\frac{1}{2}$ pounds
D. $9\frac{1}{2}$ pounds

M4D1c

18. Solve the problem below.

$$4\frac{2}{11} + 11\frac{1}{11} - 6 =$$

A. $10\frac{1}{11}$
B. $11\frac{1}{11}$
C. $9\frac{3}{11}$
D. $21\frac{1}{11}$

M4N6b

19. The cat below has turned on you. What is the number of degrees of rotation that the cat has turned?

 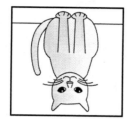

A. 120°
B. 180°
C. 360°
D. 720°

M4M2b

20. Which design below contains an isosceles triangle?

A.

B.

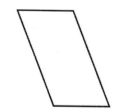

C.

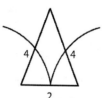

D.

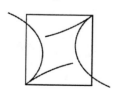

M4G1a

21. Between the numbers 12 and 4, choose the best description of their relationship.

A. 4 is one third the size of 12.

B. 12 is twice as big as 4.

C. 4 is 10 less than 12.

D. 12 is four times as big as 4.

M4A1a

22. Identify the correct operation.

$$17 \; \Box \; 47 = 64$$

A. ×

B. ÷

C. −

D. +

M4A1b

23. If you take a rectangular piece of paper 6 inches long by 3 inches high and you fold it in half, meeting the short ends, what shape do you now have?

A. A 6 inch square

B. A 3 inch square

C. A rectangle 3 inches by 6 inches

D. A trapezoid 3 inches by 6 inches

M4M2a

24. By definition, a scalene triangle has

A. two sides equal and two angles equal.

B. no sides equal and no angles equal.

C. four sides equal and all angles equal.

D. three equal sides and no angles equal.

M4G1a

25. Which object below is not a quadrilateral?

A.

B.

C.

D.

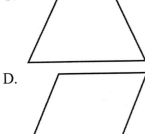

M4G1c

26. Which point on the coordinate plane below represents $(6, 3)$?

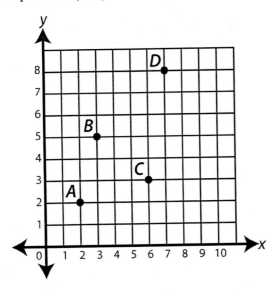

A. D
B. B
C. C
D. A

M4G3c

27. Which one answer is the correct one for both problems below?

$$2220 \div 40 =$$

$$222 \div 4 =$$

A. 55 r2

B. 56

C. 111

D. 11 r1

M4N4d

28. Using the graph below, which kind of movie was seen over Winter Break by about 350 students?

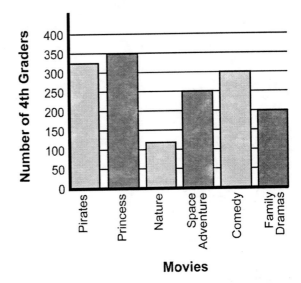

A. pirate movies

B. nature movies

C. family dramas

D. princess movies

M4D1a

29. Solve $57.6 \times 2 =$

   A. 115.2

   B. 112.5

   C. 343.2

   D. 344.2

M4N5e

30. Round the following number to the nearest thousand.

   223,657.4

   A. 223,600

   B. 223,700

   C. 224,000

   D. 223,660

M4N2a

31. Solve $46.4 \div 2 = $ _____ .

   A. 23.2

   B. 23.4

   C. 42.2

   D. 42.4

M4N5d

32. What value is the digit 4 in the following number?

   142,389

   A. hundreds

   B. ten-thousands

   C. ones

   D. one hundred-thousands

M4N1a

33. Identify the three figures shown below. Your answer should be in the same order as the pictures.

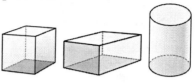

   A. square, prism, pyramid

   B. cube, prism, cylinder

   C. cylinder, prism, square

   D. cube, cylinder, prism

M4G2c

34. Which ordered pair represents the point on the coordinate plane below?

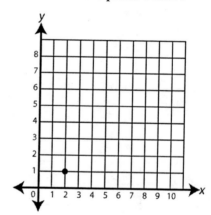

   A. $(2,0)$

   B. $(1,0)$

   C. $(1,2)$

   D. $(2,1)$

M4G3b

35. Round the following number to the nearest whole number.

   45,658.49

   A. 45,658

   B. 45,659

   C. 45,658.4

   D. 45,658.5

M4N2c

Part Two

1. Solve the following problem.

$$74 \div 4 =$$

A. 18 r3

B. 18 r2

C. 17 r3

D. 17 r2

M4N4b

2. Identify the correct operation.

$$12 \square 4 = 8$$

A. −

B. ×

C. ÷

D. +

M4A1b

3. In the illustration below, what degree of rotation did the fish make?

A. 180°

B. 360°

C. 90°

D. 45°

M4M2b

4. Round the number 8,964,499 to the nearest thousand.

A. 8,960,000

B. 8,964,500

C. 8,964,000

D. 9,000,000

M4N2a

5. $1\frac{2}{3} - \frac{1}{3} =$

A. $1\frac{1}{3}$

B. $\frac{4}{3}$

C. $1\frac{2}{3}$

D. 1

M4N6b

6. Put the following in order from greatest to least.

7.3   5.4   8.7   3.7   4.5

A. 7.3 8.7 5.4 4.5 3.7

B. 3.7 4.5 5.4 7.3 8.7

C. 8.7 5.4 7.3 4.5 3.7

D. 8.7 7.3 5.4 4.5 3.7

M4N5b

7. Which design contains a trapezoid?

A.

B.

C.

D.

M4G1c

8. Which ordered pair represents the point in the coordinate plane below?

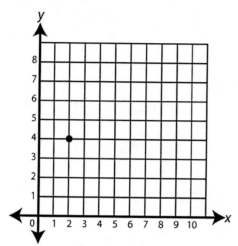

A. (2, 0)
B. (0, 4)
C. (4, 2)
D. (2, 4)

M4G3b

9. What number is one million, three hundred thousand, seventeen in standard form?

A. 1,300,017
B. 3,100,017
C. 1,317,000
D. 3,317,000

M4N1b

10. Round 12.7 to the nearest whole number.

A. 12.1
B. 10
C. 13
D. 13.1

M4N2c

11. $14.5 - 11.6 =$

A. 3.9
B. 4.5
C. 2.9
D. 3.5

M4N5c

12. Which digit is the hundreds place for 8,951,624?

A. 6
B. 2
C. 1
D. 9

M4N1a

13. Which number in the following sentence is the dividend?

$$45 \div 9 = 5$$

A. 9
B. 45
C. 5
D. =

M4N4c

14. Which set of lines below are parallel?

A.

B.

C.

D.

M4G1b

15. Solve $1 \times (15 - 12) + 5 \times (18 - 16) =$

A. 14
B. 11
C. 16
D. 13

M4N7b

Use the graph below to answer questions 16 and 17.

Mrs. Eatyore's Vegetable Garden Yield

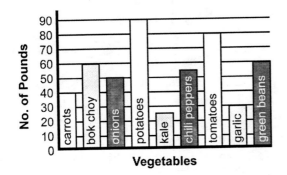

16. How many more potatoes did Mrs. Eatyore grow than garlic and kale together?

    A. 35
    B. 55
    C. 90
    D. 145

    M4D1c

17. What two vegetables did Mrs. Eatyore grow about the same amount of?

    A. kale and tomatoes
    B. potatoes and tomatoes
    C. bok choy and green beans
    D. carrots and garlic

    M4D1b

18. Divide: $62.4 \div 2 =$

    A. 31.2
    B. 34.2
    C. 41.2
    D. 31.4

    M4N5e

19. Mary's miniature pinscher, Minnie, had 4 puppies weighing 1 pound and 1 ounce, 1 pound and 2 ounces, 15 ounces, and the smallest at 14 ounces. How much in total did the puppies weigh?

    A. 3 pounds, 15 ounces
    B. 4 pounds, 1 ounce
    C. 3 pounds
    D. 4 pounds

    M4M1a

20. What is the degree of rotation of the figures below?

    A. 120°
    B. 90°
    C. 180°
    D. 360°

    M4M2b

21. What is 17,771 in word form?

    A. One hundred seven thousand, seven hundred seventeen
    B. Seventeen hundred, seven hundred seventeen
    C. Seventeen thousand, seven hundred seventy-one
    D. Seventeen thousand, seven hundred seventeen

    M4N1b

22. Suppose you were given the data and key found below. How many symbols would need to be drawn to represent winter? (Each symbol equals 10°.)

Average Temperature of Seasons

Temperature	Season
70°	spring
95°	summer
75°	fall
60°	winter

A. $9\frac{1}{2}$

B. $7\frac{1}{2}$

C. 7

D. 6

M4D1a

23. What answer solves both the equations below?

$$7,070 \div 70 =$$

$$707 \div 7 =$$

A. 101
B. 10
C. 111
D. 11

M4N4d

24. Ted's grandma is making a large bowl of potato salad. The vegetables she uses are 3 pounds of potatoes, 1 pound and 3 ounces of cucumber, 12 ounces of green onions, 8 ounces of green pepper, and 4 ounces of radishes. What is the total weight of the vegetables she uses?

A. 4 pounds and 12 ounces
B. 4 pounds and 11 ounces
C. 5 pounds and 12 ounces
D. 5 pounds and 11 ounces

M4M1c

25. Comparing the two figures below, a cube and a rectangular prism, which statement is not true?

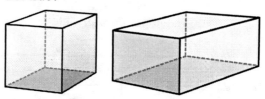

A. Both figures have twelve edges.
B. The cube is twice as big as the rectangular prism.
C. Both the figures have six faces.
D. Neither the cube or the rectangular prism have curves.

M4G2a

26. Choose the best statement to describe the relationship between 10 and 100.

A. 100 is ten times bigger than 10.

B. 100 is 80 more than 10.

C. 10 is 80 less than 100.

D. 10 is one fifth of 100.

M4A1a

27. Which statement below best describes an equilateral triangle?

A. An equilateral triangle has 4 equal sides and 4 equal angles.

B. An equilateral triangle has 2 equal sides and 3 equal angles.

C. An equilateral triangle has 3 equal sides and 3 equal angles.

D. An equilateral triangle has no equal sides and no equal angles.

M4G1a

28. Looking carefully at the diagram below, you see there is one number missing. What is the missing number?

2	4	6	8	10
3	6	9		15
4	8	12	16	20

A. 11
B. 14
C. 13
D. 12

M4D1d

29. Of the four diagrams below, which contain at least one set of parallel lines?

 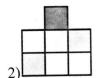

1) 2)

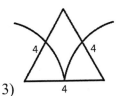

3) 4)

A. Diagram numbers 1 and 2
B. Diagram numbers 2 and 4
C. Diagram numbers 1, 2 and 4
D. Diagram numbers 2, 3 and 4

M4G1b

30. Randy had to get some new clothes for school. Randy's mom bought him 2 shirts that cost $12.00 each and 1 pair of pants that cost $14.00. Which of the following should Randy's mom give the cashier?

A. 1 ten dollar bill and 1 twenty dollar bill
B. 2 twenty dollar bills
C. 1 fifty dollar bill and 1 ten dollar bills
D. 3 five dollar bills and 3 one dollar bills

M4N2d

31. Which diagram below has both perpendicular and parallel lines like those you would find on a rectangular prism?

A.

B.

C.

D.

M4G2b

32. Solve: $23\frac{2}{3} - 11\frac{1}{3} + 12 =$

A. $24\frac{1}{3}$

B. $46\frac{2}{3}$

C. $23\frac{1}{3}$

D. $34\frac{2}{3}$

M4N6b

33. If you are trying to keep track of how much money you have going in and out of your piggy bank, which operations would you most likely use?

A. Multiplication and subtraction.

B. Division and multiplication.

C. Addition and subtraction.

D. Addition and division.

M4N7a

34. Multiply 239 by 47.

A. 11,233

B. 12,233

C. 11,237

D. 12,237

M4N3

35. Round 578,875 to the nearest thousand.

A. 578,800

B. 579,000

C. 579,800

D. 578,880

M4N2a

Practice Test 2

Part One

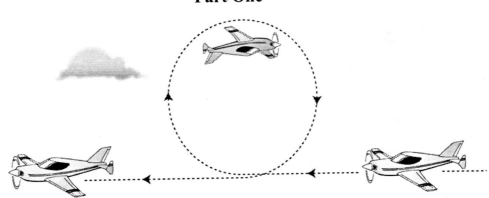

1. Andy is a small plane pilot who likes to give his riders a thrill by doing a loop-de-loop before they land. How many degrees of rotation is one full loop-de-loop?

   A. 120°

   B. 180°

   C. 360°

   D. 240°

   M4M2b

2. What is eleven thousand, two hundred eleven in standard form?

   A. 11,211

   B. 111,221

   C. 11,221

   D. 112,112

   M4N1b

3. Solve $24.8 \div 4 =$

   A. 3.2

   B. 4.6

   C. 4.8

   D. 6.2

   M4N5e

4. Solve: $4 \times (12 - 9) + 3 \times (54 - 49) =$

   A. 27

   B. 15

   C. 30

   D. 24

   M4N7b

5. Round 345,543 to the nearest ten.

   A. 345,530

   B. 345,540

   C. 345,000

   D. 345,500

   M4N2a

6. Reggie has invited 32 relatives over for a celebration. He has 67 hotdogs to share. If Reggie and all of his relatives each eat two hotdogs, how many hotdogs, if any, will be left over?

   A. 5 hotdogs are leftover

   B. 4 hotdogs are leftover

   C. 1 hotdog is leftover

   D. 0 hotdogs are leftover

   M4N4b

Use the graph below to answer questions 7 and 8.

Little League Team Wins

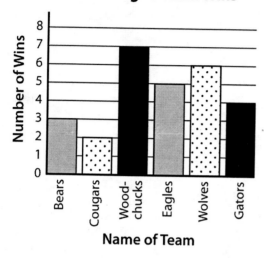

7. Which little league team won the most games?

A. Woodchucks
B. Wolves
C. Gators
D. Cougars

M4D1a

8. What is the difference in the number of wins between the Eagles and the Bears?

A. 2 wins
B. 3 wins
C. No difference
D. 4 wins

M4D1c

9. Identify the correct operation.

$$25 \square 5 = 5$$

A. ×
B. +
C. −
D. ÷

M4A1b

10. Which place value is the digit 7 in the number?

8,798,898

A. hundreds
B. hundred-thousands
C. thousands
D. ten-thousands

M4N1a

11. Put the following numbers in order from smallest to greatest.

45.26   45.21   45.27   45.3   45.11

A. 45.26   45.21   45.27   45.3   45.11
B. 45.11   45.21   45.3   45.26   45.27
C. 45.11   45.21   45.26   45.27   45.3
D. 45.21   45.11   45.26   45.27   45.3

M4N5b

12. If Carmella buys fourteen 6 ounce frozen fish burgers for her Friday fish fry, how many pounds of fish does Carmella buy?

A. five pounds, 4 ounces
B. five pounds, 6 ounces
C. four pounds, 12 ounces
D. four pounds, 15 ounces

M4M1a

13. What is the quotient in the following problem?

$$72 \div 8 = 9$$

A. 9
B. 72
C. 8
D. Does not apply

M4N4c

14. Which coordinate pair is represented by point B in the coordinate plane shown above?

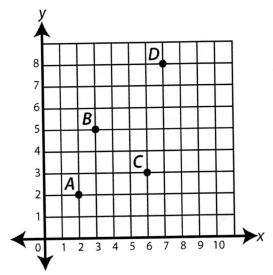

A. $(2, 2)$

B. $(5, 3)$

C. $(2, 5)$

D. $(3, 5)$

M4G3b

15. Using the graph below, what is the total number of blue, green, and black shirts worn on the first day of school?

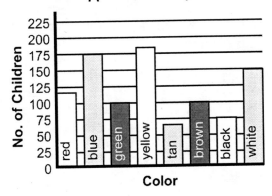

A. 325

B. 350

C. 275

D. 375

M4D1c

16. Round 467.45 to the nearest whole number.

A. 467

B. 468

C. 467.4

D. 468.45

M4N2c

17. Solve $4\frac{4}{7} - 2\frac{3}{7} =$

A. $1\frac{1}{7}$

B. $2\frac{2}{7}$

C. $2\frac{1}{7}$

D. $2\frac{6}{7}$

M4N6b

18. Which sentence below does not describe the relationship between 2 and 10?

A. 2 multiplied by $5 = 10$

B. 10 divided by $2 = 5$

C. 10 minus $2 = 5$

D. 10 is 5 times larger than 2

M4A1a

19. Which triangle does not have any angles or sides the same?

A. equilateral

B. scalene

C. isosceles

D. oddolene

M4G1a

20. Which of the following signs has at least two sets of parallel lines?

A.

B.

C.

D.

Buffle's Bakery

M4G1b

21. Brittney is buying school supplies. She is buying two packages of markers at $2.99 each, three packages of pencils at $0.99 each, and one notebook at $3.99. She wants to give the cashier an amount of money so she will have the least amount of change given back to her. Which amount should she hand the cashier?

A. 1 ten dollar bill
B. 2 ten dollar bills
C. 3 five dollar bills
D. 1 five dollar bill

M4N2d

22. $655 \times 55 =$

A. 36,025
B. 65,555
C. 35,055
D. 36,550

M4N3

23. What kind of triangle sits atop the square in the diagram below?

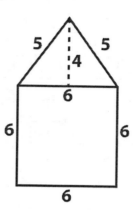

A. equilateral
B. isosceles
C. oddolene
D. scalene

M4G1a

24. Solve: $644.72 - 322.9 =$ _____.

A. 321.82
B. 322.63
C. 321.63
D. 322.82

M4N5c

25. Change the improper fraction $\frac{28}{3}$ into a mixed number.

A. $8\frac{1}{3}$

B. $8\frac{2}{3}$

C. $9\frac{1}{3}$

D. $9\frac{2}{3}$

M4N6c

26. You cut out two circles and a long rectangle out of cardboard. The two circles form the top and bottom and the rectangle becomes the middle of what object?

   A. rectangular prism

   B. cube

   C. sphere

   D. cylinder

M4G2c

27. What is the coordinate pair for the point on the coordinate plane above?

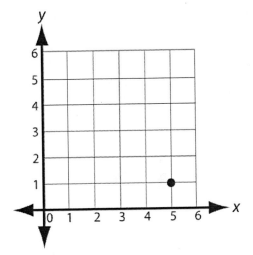

   A. $(5, 0)$

   B. $(1, 0)$

   C. $(1, 5)$

   D. $(5, 1)$

M4G3b

28. Round $452,452$ to the nearest ten.

   A. $452,450$

   B. $452,400$

   C. $452,452$

   D. $452,000$

M4N2a

29. What is eight hundred forty-three thousand, six hundred twenty-nine in standard form?

   A. $843,629$

   B. $846,329$

   C. $843,692$

   D. $843,292$

M4N1b

30. How many sides does a quadrilateral have?

   A. 3

   B. 4

   C. 8

   D. 6

M4G1c

31. Janeen glued two cubes from a game together. What object does the two glued cubes now resemble?

   A. cylinder

   B. cone

   C. rectangular prism

   D. a larger cube

M4G2c

32. The penguin has moved. What is the number of degrees of rotation that the penguin has turned?

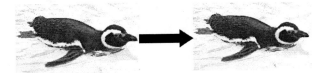

   A. $120°$

   B. $180°$

   C. $360°$

   D. $90°$

M4M2b

33. Which of the following statements describing a rectangle is false?

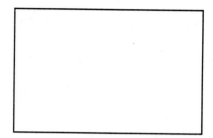

A. The rectangle has two sets of parallel sides.

B. The rectangle has four 90° angles.

C. The rectangle is longer than a square.

D. The rectangular has three equal sides and angles.

M4G2b

34. Identify the correct operation.

$$4 \triangle 44 = 176$$

A. ×

B. ÷

C. +

D. −

M4A1b

35. Francisco was pushing a wheel barrel full of 200 pounds of gravel to fill in the holes on his driveway. The first hole used $12\frac{1}{2}$ pounds of gravel, the second used 36 pounds, the third used $52\frac{1}{2}$ pounds, and the fourth used just $5\frac{1}{2}$ pounds. How much gravel did Francisco have left in his wheel barrel after filling the four holes? Express your answer in pounds and ounces.

A. 106 pounds and 8 ounces

B. 105 pounds and 8 ounces

C. 93 pounds and 8 ounces

D. 92 pounds and 8 ounces

M4M1a

Part Two

Each symbol represents 10 flowers.

1. In the picture above, each symbol represents 10 flowers in Grandmother's birthday bouquet from all her grandchildren. How many flowers in all were there in the bouquet?

   A. 110
   B. 120
   C. 13
   D. 12

M4D1a

2. Round 479 to the nearest hundred.

   A. 500
   B. 400
   C. 470
   D. 570

M4N2a

3. What is three hundred thousand, seven hundred in standard form?

   A. 370,000
   B. 300,070
   C. 300,700
   D. 373,000

M4N1b

4. $45.27 + 890.27 =$

   A. 135.54
   B. 935.45
   C. 953.54
   D. 935.54

M4N5c

5. Focus on the triangle within the figure below. What kind of triangle is it?

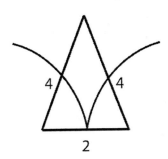

   A. scalene
   B. isosceles
   C. equilateral
   D. none of the above

M4G1a

6. Place the following in order from greatest to least.

   2.5 5.2 3.6 6.3 4.4

   A. 6.3 5.2 4.4 3.6 2.5
   B. 2.5 3.6 4.4 5.2 6.3
   C. 4.4 6.3 5.2 3.6 2.5
   D. 2.5 4.4 5.2 3.6 6.3

M4N5b

7. Multiply 223 by 99.

   A. 2,277
   B. 22,077
   C. 22,777
   D. 2,077

M4N3

8. What is the coordinate pair for the point D on the coordinate plane below?

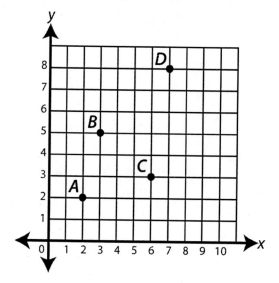

A. $(8, 7)$

B. $(8, 0)$

C. $(7, 8)$

D. $(0, 7)$

M4G3b

9. Which number is the quotient in $70 \div 10 = 7$?

A. 70

B. 10

C. 7

D. None of the above.

M4N4c

10. $10 \times (4 - 2) + 2 \times (8 - 3) + 12 =$

A. 30

B. 38

C. 57

D. 42

M4N7b

11. Aunt Mary is making purses for each of four nieces. Each purse takes $\frac{3}{4}$ of a yard of her fabric. Aunt Mary wants to add an additional pocket that will take another $\frac{1}{4}$ of a yard per purse. How many whole yards should Aunt Mary buy?

A. 4 yards

B. 3 yards

C. 5 yards

D. 2 yards

M4N6b

12. How many quadrilaterals are in the diagram below?

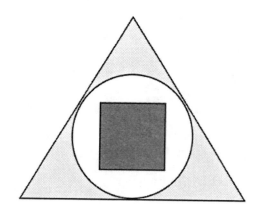

A. 2

B. 1

C. 3

D. None

M4G1c

13. What degree of rotation did the figure L make to become R?

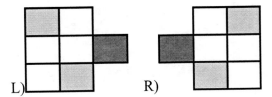

L)   R)

A. 120°

B. 90°

C. 240°

D. 180°

M4M2b

14. Identify the property of the equation.

$6 + (10 + 12) = (6 + 10) + 12$

A. Distributive
B. Commutative
C. Associative
D. None of the above.

M4N7c

15. A shoebox is shaped most like which of the following?

A. square pyramid
B. cube
C. cylinder
D. rectangular prism

M4G2c

16. Jake's dad is having a barbecue this weekend for family and friends. He buys 4 ounces of hamburger for each of 7 children and 6 ounces of chicken for each of 9 adults. How much meat in total did Jake's dad buy? Express your answer in pounds and ounces.

A. 6 pounds and 2 ounces
B. 5 pounds and 7 ounces
C. 6 pounds and 8 ounces
D. 5 pounds and 2 ounces

M4M1a

17. What value place is the digit 7 in 879,986?

A. ten-thousands
B. thousands
C. hundred-thousands
D. hundreds

M4N1a

18. Each sign below represents 5 signs that Catherine and Maureen saw while on a long car ride with their parents. How many more stop signs did the girls see than yield signs?

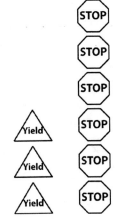

A. 30
B. 15
C. 10
D. 35

M4D1c

19. Rounding numbers can be a very useful tool when shopping. You can round up each item you buy to the nearest dollar or half dollar so you are sure you have enough money with you to pay for everything. Alisha was using this technique shopping for clothes. She wanted to buy a pair of jeans for $15.99 and a blouse for $12.59 and a scarf for $3.79. She has $30.00 with her. If she doesn't have enough money, what is the least expensive item she should leave behind?

A. Yes, she has enough.
B. No, she is short on money and should leave the jeans.
C. No, she is short on money and should leave the scarf.
D. No, she is short on money and should leave the blouse.

M4N2b

20. Which of the following are parallel lines?

A.

B.

C.

D.

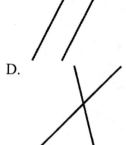

M4G1b

21. Choose the correct operation for the following problem.

$$6 \triangle 6 = 36$$

A. ×

B. +

C. −

D. ÷

M4A1b

22. The first figure below has been rotated to the second figure! How many degrees of rotation?

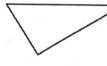

A. 360°

B. 120°

C. 90°

D. 180°

M4M2b

23. What is $\frac{4}{5} - \frac{1}{5} = ?$

A. $\frac{2}{5}$

B. $\frac{3}{5}$

C. $\frac{5}{5}$

D. $\frac{4}{5}$

M4N6b

24. What goes up, but never goes down? Age! Which operation do we use to increase our age each birthday?

A. ×

B. ÷

C. +

D. −

M4A1a

25. Round 445.51 to the nearest whole number.

A. 445.5

B. 446

C. 445

D. 446.1

M4N2c

26. Which of the angles below is required for any right triangle?

A.

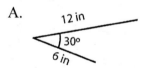

B.

C.

D.

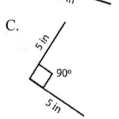

M4G1a

27. $5\frac{1}{5} + 4\frac{3}{5} =?$

   A. $10\frac{3}{5}$

   B. $9\frac{3}{5}$

   C. $10\frac{4}{5}$

   D. $9\frac{4}{5}$

   M4N6b

28. $87 \div 4 =$

   A. 20 r3

   B. 20 r2

   C. 21 r2

   D. 21 r3

   M4N4b

29. Friends on a beach decided to have a contest to see who could collect the most pounds of seashells in one hour. Pamela found 8 ounces; Sophia found 12 ounces; Jeremy found 2 pounds, 2 ounces; Cesar found 1 pound, 14 ounces; and Jemma found 1 pound, 2 ounces. What is the total weight of all the shells found by the group?

   A. 6 pounds, 12 ounces

   B. 6 pounds, 6 ounces

   C. 7 pounds, 2 ounces

   D. 6 pounds, 8 ounces

   M4M1a

30. What is 45,454 in word form?

   A. Forty five thousand, four hundred fifty-four

   B. Fifty four thousand, four hundred fifty-four

   C. Forty five thousand, five hundred forty-four

   D. Forty five thousand, four hundred forty-four

   M4N1b

31. What one answer is correct for both of the following problems?

   $3,070 \div 10 =$

   $307 \div 1 =$

   A. 371

   B. 3,060

   C. 307

   D. 306

   M4N4d

Use the graph below to answer questions 32 and 33.

Potted Plants

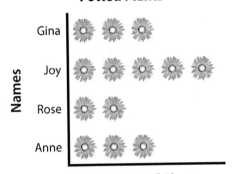

Key: 1 flower equals 10 plants

32. How many more potted plants did Joy plant than Rose?

   A. 20

   B. 30

   C. 50

   D. 10

   M4D1c

33. What is the total number of plants that Anne and Gina planted?

   A. 80

   B. 50

   C. 70

   D. 60

   M4D1c

34. Multiply $88.24 \times 2 =$

    A. 176.48

    B. 176.84

    C. 174.68

    D. 174.48

M4N5e

35. Convert $24\frac{6}{4}$ into a proper, mixed number.

    A. $24\frac{2}{4}$

    B. $25\frac{1}{4}$

    C. $25\frac{2}{4}$

    D. $24\frac{3}{4}$

M4N6c

Index

ANSWER KEY

FOR

Mastering the Georgia
4th Grade CRCT
in
Mathematics

Erica Day

Colleen Pintozzi

Mary Reagan

AMERICAN BOOK COMPANY

P. O. BOX 2638

WOODSTOCK, GA 30188-1383

TOLL FREE 1 (888) 264-5877   PHONE (770) 928-2834   FAX (770) 928-7483

Web site: www.americanbookcompany.com

The standards listed at the beginning of each chapter correspond to the following Georgia Performance Standards descriptions.

Numbers and Operations. Students will further develop their understanding of whole numbers and master the four basic operations with whole numbers by solving problems. They will understand rounding and when to appropriately use it. Students will add and subtract decimal fractions and common fractions with common denominators.

M4N1 Students will further develop their understanding of how whole numbers and decimal fractions are represented in base-ten numeration system.

(A) Identify place value names and places from hundredths through one million.

(B) Equate a number's word name, its standard form, and its expanded form.

M4N2. Students will understand and apply the concept of rounding numbers.

(A) Round numbers to the nearest ten, hundred, or thousand.

(B) Describe situations in which rounding numbers would be appropriate and determine whether to round to the nearest ten, hundred, or thousand.

(C) Understanding the meaning of rounding a decimal fraction to the nearest whole number.

(D) Represent the results of computation as a rounded number when appropriate and estimate a sum or difference by rounding numbers.

M4N3. Students will solve problems involving multiplication of 2-3 digit numbers by 1 and 2 digit numbers.

M4N4. Students will further develop their understanding of division of whole numbers and divide in problem solving situations without calculators.

(A) Know the division facts with understanding and fluency.

(B) Solve problems involving division by a 2-digit number (including those that generate a remainder).

(C) Understand the relationship between dividend, divisor, quotient, and remainder.

(D) Understand and explain the effect on the quotient of multiplying or dividing both the divisor and dividend by the same number. ($2,050 \div 50$ yields the same answer as $205 \div 5$.)

M4N5. Students will further their understanding of the meaning of decimal fractions and use them in computations.

(A) Understand decimal fractions are a part of the base-ten system.

(B) Understand the relative size of numbers and order two digit decimal fractions.

(C) Add and subtract both one and two digit decimal fractions.

(D) Model multiplication and division of decimal fractions by whole numbers.

(E) Multiply and divide both one and two digit decimal fractions by whole numbers.

M4N6. Students will further develop their understanding of the meaning of common fractions and use them in computations.

(A) Understand representations of simple equivalent fractions.

(B) Add and subtract fractions and mixed numbers with common denominators. (Denominators should not exceed twelve.)

(C) Convert and use mixed numbers and improper fractions interchangeably.

M4N7. Students will explain and use properties of the four arithmetic operations to solve and check problems.

  (A) Describe situations in which the four operations may be used and the relationships among them.
  (B) Compute using the order of operations, including parentheses.
  (C) Compute using the commutative, associative, and distributive properties.
  (D) Use mental math and estimation strategies to compute.

Measurement. Students will measure weight in appropriate metric and standard units. They will also measure angles.

M4M1. Students will understand the concept of weight and how to measure weight.

  (A) Use standard and metric units to measure the weight of objects.
  (B) Know units used to measure weight (gram, kilogram, ounces, pounds and tons).
  (C) Compare one unit to another within a single system of measurement.

M4M2. Students will understand the concept of angles and how to measure angles.

  (A) Use tools, such as a protractor or angle ruler, and other methods such as paper folding, drawing a diagonal in a square, to measure angles.
  (B) Understand the meaning and measure of a half rotation ($180°$) and a full rotation ($360°$).
  (C) Determine that the sum of the three angles in a triangle is always $180°$.

Geometry. Students will understand and construct plane and solid geometric figures. They will also graph points on the coordinate plane.

M4G1. Students will define and identify the characteristics of geometric figures through examination and construction.

  (A) Examine and compare angles in order to classify and identify triangles by their angles.
  (B) Describe parallel and perpendicular lines in plane geometric figures.
  (C) Examine and classify quadrilaterals (including parallelograms, squares, rectangles, trapezoids, and rhombi).
  (D) Compare and contrast the relationships among quadrilaterals.

M4G2. Students will understand fundamental solid figures.

  (A) Compare and contrast a cube and a rectangular prism in terms of the number and shape of their faces, edges, and vertices.
  (B) Describe parallel and perpendicular lines and planes in connection with the rectangular prism.
  (C) Construct/collect models for solid geometric figures (cubes, prisms, cylinders, pyramids, spheres, and cones).

M4G3. Students will use the coordinate system.

  (A) Understand and apply ordered pairs in the first quadrant of the coordinate system.
  (B) Locate a point in the first quadrant in the coordinate plane and name the ordered pair.
  (C) Graph ordered pairs in the first quadrant.

Algebra. Students will investigate and represent mathematical relationships between quantities using mathematical expressions in problem-solving situations.

M4A1. Students will represent and interpret mathematical relationships in quantitative expressions.

(A) Understand and apply patterns and rules to describe relationships and solve problems.
(B) Represent unknowns using symbols, such as $\triangle$ and $\square$.
(C) Write and evaluate mathematical expressions using symbols and different values.

Data Analysis. Students will gather, organize, and display data. They will also compare features of graphs.

M4D1. Students will gather, organize, and display data according to the situation and compare related features.

(A) Represent data in bar, line and pictographs.
(B) Investigate the features and tendencies of graphs.
(C) Compare different graphical representation for a given set of data.
(D) Identify missing information and duplications in data.

Process Standards. Each topic studied in this course would be developed with careful thought toward helping every student achieve the following process standards.

M4P1. Students will solve problems (using appropriate technology).

(A) Build new mathematical knowledge through problem solving.
(B) Solve problems that arise in mathematics and in other contexts.
(C) Apply and adapt a variety of appropriate strategies to solve problems.
(D) Monitor and reflect on the process of mathematical problem solving.

M4P2. Students will reason and evaluate mathematical arguments.

(A) Recognize reasoning and proof as fundamental aspects of mathematics.
(B) Make and investigate mathematical conjectures.
(C) Develop and evaluate mathematical arguments and proofs.
(D) Select and use various types of reasoning and methods of proof.

M4P3. Students will communicate mathematically.

(A) Organize and consolidate their mathematical thinking through communication.
(B) Communicate their mathematical thinking coherently and clearly to peers, teachers, and others.
(C) Analyze and evaluate the mathematical thinking and strategies of others.
(D) Use the language of mathematics to express mathematical ideas precisely.

M4P4. Students will make connections among mathematical ideas and to other disciplines.

(A) Recognize and use connections among mathematical ideas.
(B) Understand how mathematical ideas interconnect and build on one another to produce a coherent whole.
(C) Recognize and apply mathematics in contexts outside of mathematics.

M5P5. Students will represent mathematics in multiple ways.

(A) Create and use representations to organize, record, and communicate mathematical ideas.
(B) Select, apply, and translate among mathematical representations to solve problems.
(C) Use representations to model and interpret physical, social, and mathematical phenomena.

Chart of Standards

Standard	Chapter Number	Diagnostic Test Part 1 Question #	Diagnostic Test Part 2 Question #	Practice Test 1 Part 1 Question #	Practice Test 1 Part 2 Question #	Practice Test 2 Part 1 Question #	Practice Test 2 Part 2 Question #
Number and Operations							
M4N1a	1	2	1, 26	32	12	10	17
M4N1b	1	10, 22	19	1	9, 21	2, 29	3, 30
M4N2a	1	4, 26	6	4, 30	4, 35	5, 28	2
M4N2b	1						19
M4N2c	4	14	10, 30	12, 35	10	16	25
M4N2d	1	29	27	6	30	21	
M4N3	2	11	17	9	34	22	7
M4N4a	2						
M4N4b	2	15, 23	7	15	1	6	28
M4N4c	2	30		16	13	13	9
M4N4d	2		20	27	23		31
M4N5a	4						
M4N5b	4	28	8	14	6	11	6
M4N5c	4	5	29, 35	8	11	24	4
M4N5d	4			31			
M4N5e	4	34	14	29	18	3	34
M4N6a	3		21				
M4N6b	3	20, 35	2, 34	18	5, 32	17	11, 23, 27
M4N6c	3					23	35
M4N7a	5				33		
M4N7b	5	12	11	10	15	4	10
M4N7c	5						14
M4N7d	5						

Chart of Standards (continued)

Standard	Chapter Number	Diagnostic Test Part 1 Question #	Diagnostic Test Part 2 Question #	Practice Test 1 Part 1 Question #	Practice Test 1 Part 2 Question #	Practice Test 2 Part 1 Question #	Practice Test 2 Part 2 Question #
Measurement							
M4M1a	7				19	12, 35	16, 29
M4M1b	7			5			
M4M1c	7	31	12, 15		24		
M4M2a	8	3, 16	22	23			
M4M2b	8	1	3	3, 19	3, 20	1, 32	13, 22
M4M2c	9						
M4G3a	11						
M4G3b	11	7, 21, 27	5, 28	13, 34	8	14, 27	8
M4G3c	11			26			
Algebra							
M4A1a	6	33		21	26	18	24
M4A1b	6	18, 24	4, 23	2, 22	2	9, 34	21
M4A1c	6						
Data Analysis							
and Probability							
M4D1a	12	13	25	11, 28	22	7	1
M4D1b	12	19			17		
M4D1c	12	8	9, 32, 33	17	16	15	18, 32, 33
M4D1d	12				28	8	

Diagnostic Test
Part One
Pages 1–6

1. C	6. D	11. A	16. A	21. A	26. D	31. B
2. B	7. A	12. C	17. A	22. B	27. C	32. D
3. B	8. C	13. B	18. C	23. A	28. C	33. C
4. A	9. A	14. B	19. B	24. B	29. A	34. A
5. D	10. B	15. A	20. D	25. A	30. A	35. C

Diagnostic Test
Part Two
Pages 7–12

1. A	6. A	11. A	16. D	21. A	26. B	31. D
2. C	7. C	12. D	17. A	22. C	27. D	32. B
3. D	8. C	13. B	18. B	23. B	28. A	33. B
4. D	9. B	14. C	19. A	24. C	29. B	34. C
5. B	10. D	15. C	20. C	25. A	30. C	35. C

Chapter 1 Whole Numbers: Place Values and Rounding
Page 16 Place Value: Greater Than One

1. Two thousand (2,000)
2. Two hundred-thousand (200,000)
3. Two (2)
4. Two million (2,000,000)
5. Twenty (20)
6. Two hundred (200)
7. Twenty thousand (20,000)
8. Two hundred (200)
9. 2
10. 4
11. 5
12. 1
13. 3
14. 8
15. 6
16. five ones
17. three million
18. six tens
19. twenty thousands or two ten-thousands
20. six million
21. one hundred-thousand
22. eight tens or eighty
23. two tens or twenty

Page 17 Numbers in Expanded Form

1. $2 \times 100 + 4 \times 10 + 3 \times 1$

2. $6 \times 1,000 + 4 \times 100 + 5 \times 10 + 8 \times 1$

3. $1 \times 1,000,000 + 4 \times 100,000 + 4 \times 10,000 + 6 \times 1,000 + 8 \times 100 + 7 \times 10 + 9 \times 1$

4. $6 \times 100,000 + 2 \times 10,000 + 1 \times 1,000 + 3 \times 100 + 7 \times 10 + 8 \times 1$

5. $5 \times 10 + 2 \times 1$

6. $3 \times 10,000 + 6 \times 1,000 + 1 \times 100 + 1 \times 10 + 2 \times 1$

7. $1 \times 100,000 + 7 \times 10,000 + 4 \times 1,000 + 2 \times 100 + 2 \times 10 + 4 \times 1$

8. $8 \times 100 + 6 \times 10 + 2 \times 1$

9. $4 \times 10,000 + 7 \times 1,000 + 8 \times 100 + 8 \times 10 + 3 \times 1$

10. $7 \times 100 + 7 \times 10 + 7 \times 1$

11. $2 \times 10,000 + 4 \times 1,000 + 6 \times 100$

12. $9 \times 1,000,000 + 1 \times 100,000 + 1 \times 10,000 + 8 \times 1,000 + 7 \times 100 + 6 \times 10 + 6 \times 1$

Page 20 Writing Numbers in Word Form

1. four hundred fifty-one

2. two million, one hundred twelve thousand, two hundred twenty-two

3. six thousand, five hundred forty-seven

4. twenty-seven

5. eight hundred twelve thousand, three hundred twenty-one

6. fifty-six thousand, seven hundred twelve

7. three hundred sixty-seven

8. five thousand, four hundred forty-four

9. four million, eight hundred eighty seven thousand, nine hundred ninety-nine

10. nineteen thousand, one hundred twenty-five

11. eight hundred seventy-two

12. twelve

13. three thousand, three hundred thirty-three

14. seven million, one hundred thirty-four thousand, six hundred twenty-eight

15. two hundred eighty-eight

16. 822
17. 1,300,212
18. 47,902

19. 13
20. 763,917
21. 42

22. 899
23. 7,683,368
24. 716

7

Pages 22–23 Rounding Whole Numbers

1. 680	7. 4,700	13. 6,450,000	19. 1,000,000	25. $2,000.00
2. 8,730	8. 13,400	14. 4,220,000	20. 4,700,000	26. 190
3. 5,672,810	9. 8,000	15. 50,000	21. 2,000,000	27. 1,600
4. 27,860	10. 13,000	16. 110,000	22. 8,000,000	28. 200,000
5. 800	11. 3,000	17. 700,000	23. 1,000,000	29. 600
6. 2,645,300	12. 8,742,000	18. 1,300,000	24. 6,000,000	30. Yes

Chapter 1 Review
Page 24

1. eight ones
2. eight hundred-thousands
3. eight thousands

4. $400 + 80 + 2 = 482$
5. $10,000 + 4,000 + 600 + 20 + 3 = 14,623$
6. $9,000 + 800 + 80 + 7 = 9,887$

7. two hundred twenty-five
8. six thousand, four hundred fifty
9. twelve
10. ten thousand, five hundred forty-nine
11. one hundred twenty-two thousand, seven hundred twenty-nine
12. two million, three hundred forty thousand, sixty-three
13. thirty two thousand, eight hundred seventy-nine
14. five hundred eighty-three

15. 660	17. 7,324,900	19. 2	21. 3
16. 34,000	18. 880,000	20. 3	

Chapter 1 Test
Pages 25–26

1. A	3. C	5. D	7. A	9. B	11. B
2. B	4. A	6. D	8. A	10. B	12. B

Chapter 2 Multiplication and Division

Page 28 Multiplying Whole Numbers

1. 5,904	6. 1,026	11. 8,550	16. 2,856
2. 7,513	7. 368	12. 1,992	17. 7,326
3. 2,784	8. 42,432	13. 10,153	18. 1,188
4. 11,988	9. 13,490	14. 4,140	19. 8,024
5. 1,344	10. 656	15. 343	20. 2,967

Pages 29–30 Dividing Whole Numbers

1. 36	6. 59 r9	11. 23 r27	16. 15 r32
2. 151	7. 211	12. 180 r20	17. 8
3. 25 r2	8. 18 r1	13. 31	18. 42
4. 24	9. 110 r6	14. 40 r4	19. 4
5. 90 r9	10. 9 r1	15. 200	20. 36

Pages 30–31 The Parts of a Division Problem

1. 81	12. 41
2. 20	13. 304
3. 16 r2	14. 2,980
4. 2	15. 444
5. 682	16. 13
6. 6	17. 1,000
7. 100	18. 2,010
8. 888	19. 334
9. 7	20. 40
10. 8,000	21. 2
11. 200	22. 3,330

Chapter 2 Review
Page 32

1. 7,344
2. 22,325
3. 6,384
4. 13,815
5. 1,755
6. 968

7. 35,303
8. 7,161
9. 21 r6
10. 80
11. 401
12. 244,004

13. 70 r2
14. 16 r35
15. 82 r57
16. 191 r10
17. 456
18. 7

19. 2
20. 5
21. 10,000
22. 741
23. 11,294
24. 10

25. 81
26. 6,000
27. 20
28. 100
29. 2
30. 1,587

Chapter 2 Test
Page 33

1. A
2. B

3. D
4. A

5. C
6. A

7. C
8. B

9. C
10. D

11. A
12. B

Page 34 Practice and Fun!

Chapter 3 Fractions
Page 36 Fraction Equivalents

1. $\frac{4}{9}$
2. $\frac{4}{12}$ or $\frac{1}{3}$
3. $\frac{6}{16}$ or $\frac{3}{8}$
4. $\frac{5}{15}$ or $\frac{1}{3}$
5. $\frac{6}{12}$ or $\frac{1}{2}$
6. $\frac{7}{21}$ or $\frac{1}{3}$

Page 37 Adding Fractions and Mixed Numbers

1. $\frac{4}{5}$
2. $\frac{10}{12}$
3. $\frac{8}{9}$
4. $\frac{5}{5}$
5. $7\frac{7}{10}$
6. $\frac{4}{7}$
7. $7\frac{10}{11}$
8. $\frac{3}{4}$
9. $\frac{9}{12}$
10. $\frac{4}{6}$
11. $\frac{3}{5}$
12. $\frac{9}{11}$
13. $11\frac{6}{7}$
14. $\frac{9}{10}$
15. $\frac{3}{7}$
16. $\frac{5}{6}$
17. $\frac{3}{4}$
18. $\frac{9}{12}$
19. $1\frac{5}{8}$
20. $\frac{7}{10}$

Page 39 Subtracting Fractions and Mixed Numbers

1. $\frac{5}{9}$
2. $\frac{7}{12}$
3. $\frac{5}{8}$
4. $3\frac{2}{7}$
5. $\frac{4}{10}$
6. $\frac{2}{6}$
7. $\frac{4}{11}$
8. $\frac{2}{5}$
9. $3\frac{1}{4}$
10. $\frac{1}{3}$
11. $6\frac{1}{12}$
12. $\frac{2}{7}$
13. $\frac{4}{9}$
14. $\frac{2}{8}$
15. $\frac{3}{10}$
16. $\frac{3}{4}$
17. $2\frac{1}{4}$
18. $\frac{3}{6}$
19. $\frac{3}{10}$
20. $\frac{4}{9}$

Page 40 Changing Mixed Numbers to Improper Fractions

1. $\frac{37}{5}$
2. $\frac{13}{4}$
3. $\frac{32}{7}$
4. $\frac{26}{5}$
5. $\frac{68}{6}$
6. $\frac{23}{3}$
7. $\frac{19}{2}$
8. $\frac{27}{6}$
9. $\frac{59}{8}$
10. $\frac{34}{5}$
11. $\frac{10}{4}$
12. $\frac{54}{7}$
13. $\frac{19}{8}$
14. $\frac{97}{9}$
15. $\frac{22}{7}$
16. $\frac{19}{4}$
17. $\frac{17}{3}$
18. $\frac{83}{7}$
19. $\frac{13}{5}$
20. $\frac{37}{3}$
21. $\frac{6}{1}$
22. $\frac{12}{1}$
23. $\frac{5}{1}$
24. $\frac{3}{1}$
25. $\frac{10}{1}$
26. $\frac{4}{1}$
27. $\frac{1}{1}$
28. $\frac{2}{1}$
29. $\frac{7}{1}$
30. $\frac{11}{1}$

Page 41 Changing Improper Fractions to Mixed Numbers

1. $2\frac{1}{8}$
2. $4\frac{3}{5}$
3. $5\frac{1}{2}$
4. $1\frac{5}{8}$
5. $4\frac{1}{6}$
6. $3\frac{1}{11}$
7. $2\frac{1}{4}$
8. $3\frac{2}{9}$
9. $2\frac{1}{3}$
10. $3\frac{3}{4}$
11. $3\frac{5}{6}$
12. $3\frac{1}{2}$
13. $9\frac{1}{9}$
14. $6\frac{2}{6}$
15. $4\frac{1}{7}$
16. $1\frac{3}{11}$
17. $8\frac{3}{8}$
18. $3\frac{2}{5}$
19. $5\frac{2}{3}$
20. $2\frac{6}{10}$

Chapter 3 Review
Page 42

1. $\frac{5}{11}$
2. $\frac{5}{11}$
3. $\frac{8}{14}$
4. $\frac{11}{12}$
5. $\frac{5}{7}$
6. $6\frac{5}{6}$
7. $\frac{9}{12}$
8. $\frac{9}{11}$
9. $9\frac{2}{3}$
10. $\frac{6}{12}$
11. $\frac{3}{10}$
12. $3\frac{4}{8}$
13. $\frac{3}{6}$
14. $\frac{4}{11}$
15. $6\frac{2}{7}$
16. $\frac{2}{4}$
17. $\frac{2}{5}$
18. $\frac{20}{3}$
19. $\frac{19}{7}$
20. $\frac{19}{4}$
21. $\frac{11}{2}$
22. $3\frac{2}{3}$
23. $3\frac{2}{5}$
24. $5\frac{2}{4}$
25. 7

Chapter 3 Test
Pages 43–45

| 1. A | 3. D | 5. B | 7. A | 9. A | 11. D | 13. A | 15. C | 17. D | 19. D |
| 2. C | 4. A | 6. C | 8. D | 10. C | 12. C | 14. B | 16. A | 18. B | 20. C |

Chapter 4 Decimals
Page 48 Reviewing Decimal Fractions

1. hundredths

2. tenths

3. tenths

4. tenths

5. hundredths

6. tenths

7. hundredths

8. tenths

9. hundredths

10. $0.2 + 0.05$

11. $10 + 0.6 + 0.08$

12. $4 + 0.1$

13. $0.1 + 0.04$

14. $0.7 + 0.03$

15. $2 + 0.1 + 0.06$

16. fourteen hundredths

17. eight and seventy-one hundredths

18. six and sixty-four hundredths

19. four dollars and ninety-nine cents

20. thirty-eight hundredths

Page 49 Ordering Decimals

1. 0.02, 0.03, 0.32, 0.33

2. 0.14, 0.18, 0.4, 0.48

3. 0.01, 0.09, 0.1, 0.11

4. 0.12, 0.3, 0.59, 0.64

5. 0.06, 0.27, 0.72, 0.88

6. 0.15, 0.21, 0.51, 0.55

7. 0.04, 0.09, 0.4, 0.9

8. 0.11, 0.15, 0.17, 0.19

9. 0.09, 0.19, 0.91, 0.99

10. 0.44, 0.47, 0.74, 0.77

11. 0.02, 0.03, 0.06, 0.62

12. 0.41, 0.43, 0.45, 0.49

13. 0.07, 0.08, 0.09, 0.10

14. 0.05, 0.15, 0.50, 0.51

15. 0.72, 0.27, 0.13, 0.12

16. 0.99, 0.88, 0.42, 0.24

17. 0.36, 0.25, 0.13, 0.01

18. 0.86, 0.68, 0.52, 0.25

19. 0.82, 0.81, 0.28, 0.18

20. 0.54, 0.5, 0.45, 0.05

Page 50 Adding Decimals

1. 14.54

2. 6.21

3. 4.49

4. 7.98

5. 8.34

6. 6.56

7. 1.25

8. 4.82

9. 2.96

10. 8.76

11. 0.66

12. 7.65

Page 51 Subtracting Decimals

1. 9.06
2. 5.44
3. 0.01
4. 0.84
5. 5.45
6. 1.89
7. 31.55
8. 1.58
9. 9.09
10. 9.99
11. 58.9
12. 10.95

Page 52 Multiplying Decimals by Whole Numbers

1. 16.1
2. 27.28
3. 57.78
4. 362.96
5. 1,160.71
6. 87.6
7. 1,602.15
8. 327.24
9. 1,676.48
10. 451.4
11. 29.91
12. 5,593.23
13. 1,331.52
14. 420.29
15. 4,162.48
16. 3,789.0
17. 796.4
18. 288.6

Page 53 Division of Decimals by Whole Numbers

1. 6.18
2. 11.1
3. 3.01
4. 21.1
5. 10.01
6. 9.1
7. 11.03
8. 100.05
9. 80.1
10. 12.73
11. 213.13
12. 4.1
13. 8.01
14. 3.02
15. 111.11
16. 11.01

Page 54 Estimating Decimals Using Rounding

1. 11
2. 279
3. 56
4. 13
5. 1,459
6. 24
7. 10
8. 147
9. 83
10. 348
11. 984
12. 4,657
13. 490
14. 753
15. 16
16. 87
17. 33
18. 47
19. 121
20. 17
21. 7,989
22. 8,251
23. 78
24. 22
25. 65
26. 82
27. 94
28. 336
29. 643
30. 145

Page 55 Decimal Word Problems

1. $39.98
2. $26.24
3. 26 pounds
4. 4.32 pounds
5. $81.06
6. $43.75

Chapter 4 Review
Page 56

1. tenths	15. 8.94	29. 10.84
2. hundredths	16. 7.52	30. 134.43
3. ones	17. 1.46	31. 4.02
4. hundredths	18. 6.27	32. 3.03
5. tenths	19. 8.96	33. 14.1
6. hundredths	20. 14.65	34. 2.07
7. $10 + 2 + 0.7$	21. 45.08	35. 8.04
8. $6 + 0.5 + 0.04$	22. 28.16	36. 6.08
9. $2 + 0.02$	23. 33.33	37. 63
10. 0.07, 0.27, 0.77, 5.27	24. 6.15	38. 83
11. 4.48, 4.84, 7.08, 8.45	25. 12.6	39. 1,665
12. 0.03, 0.3, 0.33, 3.33	26. 265.2	40. 63
13. 6.95	27. 13.04	41. 8,991
14. 7.02	28. 24.99	42. 6

Chapter 4 Test
Pages 57–58

1. D	5. A	9. D	13. A	17. B
2. A	6. D	10. A	14. D	18. A
3. B	7. C	11. B	15. B	19. B
4. C	8. B	12. C	16. C	20. B

Chapter 5 Arithmetic Operations

Page 60 Using Addition, Subtraction, Multiplication and Division

1. $3.75	2. $5.00	3. 63	4. 96	5. 114	6. $2.20

Page 62 Order of Operations

1. 1	3. 5	5. 72	7. 90	9. 20	11. 66	13. 9	15. 9	17. 32	19. 11
2. 22	4. 5	6. 28	8. 23	10. 30	12. 40	14. 2	16. 12	18. 27	20. 35

Page 63 Properties of Mathematics

1. D   2. C   3. A   4. C   5. D   6. C   7. A   8. C   9. A   10. C

Page 64 Using Mental Math and Estimation to Solve

1. 12 rolls for $10.99   2. about $48.00   3. no

Chapter 5 Review
Page 65

1. 32	6. 44	11. 14	16. 18	21. Allgood Shoe Store
2. 2	7. 6	12. 7	17. C	
3. 8	8. 20	13. 36	18. A	22. Jamie
4. 13	9. 17	14. 28	19. D	
5. 11	10. 17	15. 19	20. no	

Chapter 5 Test
Pages 66–68

1. C   3. B   5. B   7. D   9. B   11. C   13. B   15. A   17. A   19. A
2. A   4. A   6. C   8. A   10. D   12. A   14. A   16. C   18. A   20. A

Chapter 6 Mathematical Relationships in Algebra
Page 70 Patterns

1.   2.   3.   4.   5.

Page 71 Patterns Using the Four Mathematical Operations

1. 8	5. 48	9. 50	13. 77	17. 435
2. 60	6. 33	10. 49	14. 180	18. 102
3. 69	7. 69	11. 52	15. 735	19. 10
4. 9	8. 27	12. 150	16. 118	20. 392

Page 72 Patterns Using the Four Mathematical Operations

1. 5	5. 10	9. 24	13. 49	17. 50
2. 16	6. 121	10. 36	14. 60	18. 24
3. 10	7. 250	11. 48	15. 81	19. 120
4. 48	8. 1,600	12. 40	16. 1	20. 25

Page 74 Open Number Sentences

1. 3	7. 12	13. $\triangle = 3, \square = 6$	19. 913
2. 100	8. 9	14. 720	20. 7
3. 50	9. 6	15. 40	21. 6
4. 34	10. 128	16. 35	22. 555
5. 44	11. 3	17. $\square = 12, \triangle = 48$	23. 11
6. 142	12. 4	18. 6	24. $\triangle = 16, \square = 1$

Chapter 6 Review
Page 75

1.    2.    3.

4. 16	8. 3	12. 44	16. 100	20. 5
5. 64	9. 35	13. 870	17. 49	21. 111
6. 4	10. 61	14. 4	18. 4	22. 25
7. 81	11. 32	15. 2	19. 56	23. 4

Chapter 6 Test
Pages 76–78

1. C	4. A	7. C	10. D	13. C	16. B	19. B	22. C	25. C
2. A	5. C	8. B	11. B	14. B	17. A	20. D	23. A	26. B
3. D	6. D	9. A	12. A	15. A	18. D	21. A	24. A	27. D

Chapter 7 Measurement
Page 80 Measuring Weight Using Standard Units

Item	Mass in Oz.	Mass in lb.	Mass in tons
1 bottle of popcorn seeds	16 oz	1 lb.	0.0005 tons
2 packages of butter	32 oz	2 lb.	0.001 tons
1 medium elephant	64,000 oz	4,000 lb.	2 tons
1 small car	16,000 oz	1,000 lb.	0.5 tons
2 bags of candy	24 oz	1.5 lb.	0.00075 tons
1 railroad car full of cows	96,000 oz	6,000 lb.	3 tons
1 full grown cat	112 oz	7 lb.	0.0035 tons
1 large pickup truck	32,000 oz	2,000 lb.	1 ton
1 large bottle of cinnamon	8 oz	0.5 lb.	0.00025 tons

1. tons

2. ounces

3. pounds and ounces

4. pounds and ounces

5. tons

6. ounces

7. pounds

8. ounces

Page 82 Measuring Mass Using Metric Measurement

Item	Mass in grams	Mass in kilograms
2 packages of butter	1,000 g	1 kg
1 full grown cat	3,500 g	3.5 kg
1 large container of beads	500 g	0.5 kg
1 bag of apples	2,000 g	2 kg
1 small car	500,000 g	500 kg
1 full grown chicken	2,500 g	2.5 kg
4 Labrador puppies	6,000 g	6 kg
1 box of brown sugar	1,000 g	1 kg
1 big box of stuffed animals	10,000 g	10 kg
1 bag of candy	250 g	0.25 kg

1. kilograms

2. grams

3. grams

4. kilograms

5. grams

6. kilograms

7. kilograms

8. grams

Chapter 7 Review
Page 83

1. ounces
2. tons
3. tons
4. ounces
5. pounds
6. tons
7. ounces
8. pounds
9. 16
10. 2,000
11. 48
12. 4
13. 2
14. grams
15. kilograms
16. kilograms
17. kilograms
18. grams
19. kilograms
20. grams
21. kilograms
22. 1,000
23. 4,000
24. 7

Chapter 7 Test
Pages 84–85

1. C
2. B
3. A
4. B
5. B
6. D
7. C
8. A
9. C
10. B
11. C
12. D
13. B
14. A
15. D

Chapter 8 Angles and Rotation
Page 87 Angle Measurement

1. 150°; obtuse
2. 10°; acute
3. 135°; obtuse
4. 30°; acute
5. 120°; obtuse
6. 90°; right
7. 180°; straight

Page 90 Rotation

1. 180°
2. 360°
3. 180°
4. 180°

Chapter 8 Review
Pages 91–93

1. 135°; obtuse
2. 90°; right
3. 45°; acute
4. 112°; obtuse
5. 180°
6. 360°
7. 180°
8. 180°
9. 180°

Chapter 8 Test
Page 94

1. B   3. A   5. B   7. B

2. C   4. D   6. B   8. D

Chapter 9 Geometry
Page 96 Identifying Triangles By Their Angles

1. equilateral   4. isosceles   7. equilateral   10. $45°$

2. acute   5. right   8. obtuse   11. $20°$

3. right   6. right   9. $60°$ 12. $90°$

Page 98 Parallel and Perpendicular Lines

1. perpendicular   3. perpendicular and parallel   5. parallel and perpendicular

2. parallel   4. perpendicular   6. perpendicular

Page 100 Quadrilaterals

1. parallelogram   3. trapezoid   5. rhombus   7. parallelogram

2. rectangle   4. trapezoid   6. square   8. trapezoid

Chapter 9 Review
Page 101

1. acute   5. obtuse   9. parallel   13. rectangle

2. right   6. right   10. perpendicular   14. parallelogram

3. equilateral   7. perpendicular and parallel   11. trapezoid   15. false

4. isosceles   8. parallel   12. rhombus   16. true

Chapter 9 Test
Pages 102–103

1. B   3. A   5. B   7. C   9. C   11. D   13. B

2. A   4. D   6. B   8. A   10. A   12. A   14. A

Chapter 10 Solid Geometry
Page 105 Cubes and Rectangular Prisms

1. rectangular prism
2. cube
3. rectangular prism

4. cube
5. rectangular prism
6. rectangular prism
7. cube

Page 107 Solid Geometric Figures

1. rectangular prism
2. sphere
3. cone
4. cube
5. cylinder

6. pyramid
7. cube
8. cylinder
9. square pyramid
10. rectangular prism

Chapter 10 Review
Page 108

1. cone
2. cube
3. cylinder
4. square pyramid

5. rectangular prism
6. sphere
7. cylinder
8. sphere

9. cube
10. rectangular prism
11. square pyramid
12. sphere

Chapter 10 Test
Pages 109–110

1. A	3. B	5. B	7. A	9. B	11. C
2. C	4. D	6. D	8. C	10. B	12. D

Chapter 11 Coordinate Systems
Page 112 Defining a Coordinate System and Applying Ordered Pairs

1. $(3, 2)$
2. $(2, 5)$
3. $(4, 2)$
4. $(5, 1)$

Page 113–114 Graphing Points on a Coordinate Plane

1.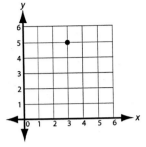

5.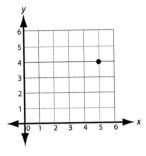

9.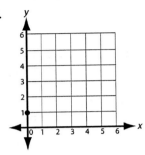

2.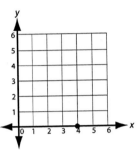

6.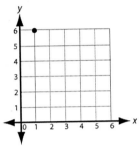

10.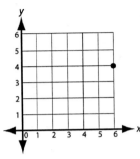

3.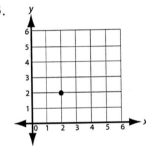

7.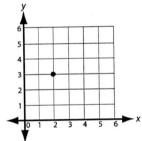

11.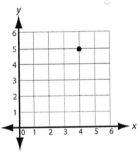

4.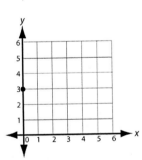

8.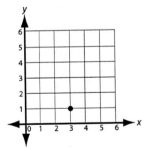

12.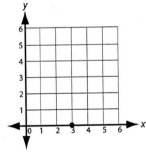

Chapter 11 Review
Page 115

1. $(2, 4)$ 2. $(6, 2)$ 3. $(3, 3)$ 4. $(5, 1)$

5.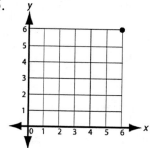

7.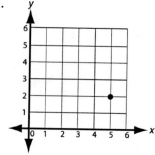

6.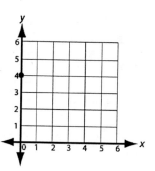

8.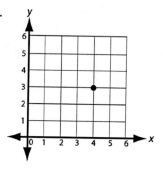

Chapter 11 Test
Pages 116–117

1. B

2. C

3. D

4. C

5. A

6. D

7. C

8. A

9. C

10. B

11. C

12. D

13. B

14. A

Chapter 12 Data Analysis
Pages 119–120 Bar Graphs, Line Graphs, and Pictographs

1. about 125

2. 150

3. 175

4. lightning bugs

5. about 575

6. about 725

7. 700

8. about 350

9. $25.00

10. $25.00

11. $25.00

12. $5.00

Page 122 Making Graphs

Little League Team Wins

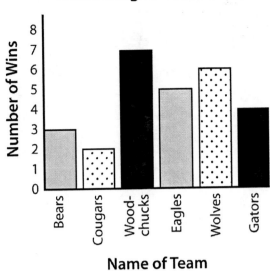

Name of Team

Page 124 Problems With Data and Graphs

1. Missing data for September.

2. Survey says 1,000 shoppers were polled, yet only 950 responses are graphed.

Chapter 12 Review
Pages 125–127

1. 150	3. 200	5. 50	7. 350
2. 50	4. 50	6. 175	8. 150

9. Missing data for the month of May.

10. 700

11. 1,700

12. July

13. January, November, and December

Chapter 12 Test
Pages 128–132

1. B	4. C	7. D	10. C	13. A
2. A	5. A	8. D	11. A	14. D
3. D	6. C	9. B	12. C	15. B

Practice Test 1
Part One
Pages 133–138

1. B	6. B	11. D	16. D	21. A	26. C	31. A
2. A	7. A	12. C	17. C	22. D	27. A	32. B
3. D	8. C	13. C	18. C	23. B	28. D	33. B
4. D	9. A	14. B	19. B	24. B	29. A	34. D
5. C	10. B	15. A	20. C	25. B	30. C	35. A

Practice Test 1
Part Two
Pages 139–144

1. B	6. D	11. C	16. A	21. C	26. A	31. D
2. A	7. B	12. A	17. C	22. D	27. C	32. A
3. B	8. D	13. B	18. A	23. A	28. D	33. C
4. C	9. A	14. B	19. D	24. D	29. A	34. A
5. A	10. C	15. D	20. C	25. B	30. B	35. B

Practice Test 2
Part One
Pages 145–150

1. C	6. C	11. C	16. A	21. C	26. D	31. C
2. A	7. A	12. A	17. C	22. A	27. D	32. C
3. D	8. A	13. A	18. C	23. B	28. A	33. D
4. A	9. D	14. D	19. B	24. A	29. A	34. A
5. B	10. B	15. B	20. C	25. C	30. B	35. C

Practice Test 2
Part Two
Pages 151–156

1. B	5. B	9. C	13. D	17. A	21. A	25. B	29. B	33. D
2. A	6. A	10. D	14. C	18. B	22. D	26. C	30. A	34. A
3. C	7. B	11. A	15. D	19. C	23. B	27. D	31. C	35. C
4. D	8. C	12. B	16. D	20. C	24. C	28. D	32. B	

Product Order Form

Please fill this form out completely and fax it to 1-866-827-3240.

American Book Company
Meeting Standards, Exceeding Expectations

Purchase Order #: _____ Date: _____

Contact Person: _____

School Name (and District, if any): _____

Billing Address: _____ Street Address: ☐ same as billing

_____ _____

Attn: _____ Attn: _____

_____ _____

_____ _____

Phone: _____ E-Mail: _____

Credit Card #: _____ Exp Date: _____

Authorized Signature: _____

Order Number	Product Title	Pricing* 10 books	Qty	Pricing 30+ books	Qty	Total Cost
GA3-M0607	Mastering the Georgia 3rd Grade CRCT in Math	$169.90 (1 set of 10 books)		$329.70 (1 set of 30 books)		
GA3-R0607	Mastering the Georgia 3rd Grade CRCT in Reading	$169.90 (1 set of 10 books)		$329.70 (1 set of 30 books)		
GA3-S0508	Mastering the Georgia 3rd Grade CRCT in Science	$169.90 (1 set of 10 books)		$329.70 (1 set of 30 books)		
GA3-H1008	Mastering the Georgia 3rd Grade CRCT in Social Studies	$169.90 (1 set of 10 books)		$329.70 (1 set of 30 books)		
GA4-M0807	Mastering the Georgia 4th Grade CRCT in Math	$169.90 (1 set of 10 books)		$329.70 (1 set of 30 books)		
GA4-R0807	Mastering the Georgia 4th Grade CRCT in Reading	$169.90 (1 set of 10 books)		$329.70 (1 set of 30 books)		
GA4-S0708	Mastering the Georgia 4th Grade CRCT in Science	$169.90 (1 set of 10 books)		$329.70 (1 set of 30 books)		
GA4-H1008	Mastering the Georgia 4th Grade CRCT in Social Studies	$169.90 (1 set of 10 books)		$329.70 (1 set of 30 books)		
GA5-M0806	Mastering the Georgia 5th Grade CRCT in Math	$169.90 (1 set of 10 books)		$329.70 (1 set of 30 books)		
GA5-R1206	Mastering the Georgia 5th Grade CRCT in Reading	$169.90 (1 set of 10 books)		$329.70 (1 set of 30 books)		
GA5-S1107	Mastering the Georgia 5th Grade CRCT in Science	$169.90 (1 set of 10 books)		$329.70 (1 set of 30 books)		
GA5-H1008	Mastering the Georgia 5th Grade CRCT in Social Studies	$169.90 (1 set of 10 books)		$329.70 (1 set of 30 books)		
GA6-L0508	Mastering the Georgia 6th Grade CRCT in ELA	$169.90 (1 set of 10 books)		$329.70 (1 set of 30 books)		
GA6-M0305	Mastering the Georgia 6th Grade CRCT in Math	$169.90 (1 set of 10 books)		$329.70 (1 set of 30 books)		
GA6-R0108	Mastering the Georgia 6th Grade CRCT in Reading	$169.90 (1 set of 10 books)		$329.70 (1 set of 30 books)		
GA6-S1206	Mastering the Georgia 6th Grade CRCT in Science	$169.90 (1 set of 10 books)		$329.70 (1 set of 30 books)		
GA6-H0208	Mastering the Georgia 6th Grade CRCT in Social Studies	$169.90 (1 set of 10 books)		$329.70 (1 set of 30 books)		
GA7-L0507	Mastering the Georgia 7th Grade CRCT in ELA	$169.90 (1 set of 10 books)		$329.70 (1 set of 30 books)		
GA7-M0305	Mastering the Georgia 7th Grade CRCT in Math	$169.90 (1 set of 10 books)		$329.70 (1 set of 30 books)		
GA7-R0707	Mastering the Georgia 7th Grade CRCT in Reading	$169.90 (1 set of 10 books)		$329.70 (1 set of 30 books)		
GA7-S1206	Mastering the Georgia 7th Grade CRCT in Science	$169.90 (1 set of 10 books)		$329.70 (1 set of 30 books)		
GA7-H0208	Mastering the Georgia 7th Grade CRCT in Social Studies	$169.90 (1 set of 10 books)		$329.70 (1 set of 30 books)		
GA8-M0305	Passing the Georgia 8th Grade CRCT in Math	$169.90 (1 set of 10 books)		$329.70 (1 set of 30 books)		
GA8-L0505	Passing the Georgia 8th Grade CRCT in Language Arts	$169.90 (1 set of 10 books)		$329.70 (1 set of 30 books)		
GA8-R0505	Passing the Georgia 8th Grade CRCT in Reading	$169.90 (1 set of 10 books)		$329.70 (1 set of 30 books)		
GA8-S0707	Mastering the Georgia 8th Grade CRCT in Science	$169.90 (1 set of 10 books)		$329.70 (1 set of 30 books)		
GA8-H0607	Mastering the Georgia 8th Grade CRCT in GA Studies	$169.90 (1 set of 10 books)		$329.70 (1 set of 30 books)		
GA8-W0907	Passing the Georgia 8th Grade Writing Assessment	$169.90 (1 set of 10 books)		$329.70 (1 set of 30 books)		

10-1-08 *Minimum order is 1 set of 10 books of the same subject.

Subtotal _____

Shipping & Handling 12% _____

American Book Company ● PO Box 2638 ● Woodstock, GA 30188-1383
Toll-Free Phone: 1-888-264-5877 ● Toll-Free Fax: 1-866-827-3240 ● Web Site: www.americanbookcompany.com

Total _____